IMPACT WEAPON COMBATIVES
BY W. HOCK HOCHHEIM

Hardcover ISBN: 978-1-932113-84-6
Paperback ISBN: 978-1-932113-44-0
Kindle ISBN: 978-1-932113-85-3

Copyright: First Edition 2011, Second Edition 2022
Published by Lauric Enterprises, Inc.
www.ForceNecessary.com
McKinney, Texas

All rights reserved.

The author, publisher, or seller do not condone or support terrorism anywhere, in any form. They do not assume any responsibility for the use or misuse of information contained in this book. The purpose of this book is to historically and artistically preserve the information contained within these pages for posterity. The book provides information that describes various methods of self-defense that may be employed against illegal aggression. In some cases individuals may make the choice to use their impact weapon to save their life or the lives of others when it is morally, legally and ethically appropriate to do so. Providing the information about how to do this in no way condones or suggests that using an impact weapon is ever justified. That decision is left completely to the individual and situation. Anyone who uses the techniques bears the responsibility entirely for any and all legal consequences of their independent actions.

Other Titles by W. Hock Hochheim
The Great Escapes of Pancho Villa
Fightin' Words
Impact Weapon Combatives
My Gun is My Passport
Last of the Gunmen
Rio Grande Black Magic
American Medieval
China Alamo
Be Bad Now
Blood Rust
Training Mission Series

IMPACT WEAPON COMBATIVES
TABLE OF CONTENTS

Prologue:	Page 5
Chapter 1: What is Force Necessary: Stick?	Page 9
Chapter 2: Introduction to Stop 6, A Primer	Page 11
Chapter 3: The Who, What, Where, When, How and Why Questions	Page 17
Chapter 4: Selecting Your Impact Weapon	Page 27
Chapter 5: Drawing Your Impact Weapon	Page 29
Chapter 6: The Impact Weapon Grips	Page 38
Chapter 7: The Impact Weapon Stances	Page 41
Chapter 8: The Combat Clock	Page 45
Chapter 9: Introduction to Footwork	Page 51
Chapter 10: Impact Weapon Strikes	Page 61
Chapter 11: Impact Weapon Blocking	Page 105
Chapter 12: While Holding. Your Support Striking and Kicking	Page 119
Chapter 13: Your Impact Weapon Total Combination Workout list	Page 129
Chapter 14: Impact Weapon Retention	Page 133
Chapter 15: The Stick Duel!	Page 151
Chapter 16: Impact Weapon Takedowns	Page 165
Chapter 17: Impact Weapons on the Ground	Page 205
Chapter 18: Test Requirements	Page 215
Addendum: The War Post	Page 225

Prologue Understanding Public Perception When Using Impact Weapons

You have drawn your impact weapon. You have done this to create a command presence to stop violence before it happens, or to stop violence while it is happening to you or someone else.

Reason to Draw 1: Stop violence before it happens.
Reason to Draw 2: Stop violence while it is happening.

People around you will be witnesses either for or against your actions, and all will be viewing the events through their own pre-conceived prejudices and memories. Many legal experts say that, while you are in action, you should proclaim aloud such words and phrases as:

"Stop!" or, "Stop! Do not make me hurt you."
"I do not want to hurt you."

In some nations, baton use is a common event, and its use goes unchallenged by the local governments. In the lower left, a modern stick vs. spray fight is about to erupt!

If you are a true, mature practitioner of modern survival strategies, then you must recognize all the pragmatic challenges that arise. You must train for the fight, survive the fight itself and survive the aftermath legally, emotionally and physically. All your actions must be appropriate, legal and moral with ethical use of force considerations of the situation. All use of force and rules of engagement must be studied and understood.

Ancient Egyptian Hieroglyphs depict fighting with sticks.

From Oakland, CA, the USA Pacific Northwest, to the Middle East, anyone traveling around the world might find themselves in a disturbance or a riot in which impact weapons are brandished. Some are dropped and recovered in the chaos. Sticks are easily bought at the ubiquitous lumber yard and stores.

This is an old newspaper photo from South America. How did all these people get expandable batons and sticks? Were these plainclothes people working undercover? Who is fighting in the forefront with the baton in hand? Good guy? Bad guy?

In China, a man fights off a wild boar with a stick.

In Southeastern US, a man charges after a local TV news reporter and cameraman.

"You have a moral, ethical and legal responsibility when selecting, carrying and using an impact weapon." - Hock

www.ForceNecessary.com

Chapter 1: What is Force Necessary: Stick?

Force Necessary: Stick is both a single-handed and double-handed tactical impact weapon system. This is taught in an integrated format because the practitioner must be able to transition to both hand grips as needed while in conflict. Fundamentally, it is about you blocking, striking and grappling while holding an impact weapon. The tactics are inspired from police, military, Spanish and Filipino methods.

Single-hand grip tactics.
Double-hand grip tactics.

Force Necessary: Stick is developed for the seamless application of all single-handed and double-handed grip impact weapon tactics in all ranges of standing, kneeling and ground fighting. This covers the entire spectrum of conflict from visual presence to deadly force as legally required. This is structured to educate the practitioner in visual, audible and physical impact weapon tactics and strategies through the following situations. The course covers:

- Stick versus unarmed attack tactics.
- Stick versus stick attack tactics.
- Stick versus knife attack tactics.
- Stick versus gun-threat attack tactics.
- Concealment/carry impact weapon strategies.
- Impact weapon quick draws, presentations and command presence.
- Impact weapon disarming and retention - counters to disarms.
- Impact weapon tactics and strategies to move and maneuver opposing personnel.
- Impact weapon tactics and strategies to control and contain opposing personnel.
- Impact weapon combat scenarios.
- Impact weapon ground fighting.
- Less-than-lethal impact weapon tactics and strategies.
- Impact weapon strategies to fight against deadly threats.

FN: Stick impact weapons cover:
  The martial arts stick or cane.
  The enforcement night stick.
  The walking and/or hiking cane.
  The long, tactical flashlight.
  The expandable baton.
  The riot baton.
  The umbrella.

The material finds its sources in:
  Military impact weapons
  Military long gun/bayonet
  Law enforcement impact weapons
  Security impact weapons
  Pacific island martial arts
    – Japan
    – Philippines
    – Indonesia
    – Hawaii

  Hybrid martial systems

A police stick.   *A common cane.*

A decent sized flashlight.   *Unique billy clubs.*

Despite the multiple influences this book will only concern itself with tactical, practical, pragmatic and event-based methods. This book will not cover the plethora of martial arts covering stick versus stick. Stick versus stick events are unlikely to occur, though we will cover the basics of such an attack because it has and might occur.

Stick versus stick disarming is another least likely event. We will at times utilize some stick versus stick in training drills where power strikes are useful, such as in blocking drills. Numerous martial arts "angles of attack" systems will be replaced here by the unforgettable, simple, common clock numbers of the *Combat Clock*.

Chapter 2: Introduction to The Stop 6: A Primer

We begin with the hand, stick, knife and gun *Stop 6* overall concept of which the stick is an integral part. They are the 6 collisions, the six common stopping points, certainly in a two-person encounter of any sort, but also in larger group confrontations.

Even if you are inside a hostile group, slivers of that group often fight one-on-one, or one versus two within the group fight, or "riot."

These are not about ranges, these are verbal and physical collisions. They are about situations and situational fighting. Training in the six situations prepare course doctrine to be thorough and complete. The less you are trained, the more you will become stuck in any one of these situational collisions.

The six stopping points include strategies and tactics with empty hand, the impact weapon-"stick," the knife and the gun. The Stop 6 format is just a laboratory that automatically covers most fighting problems.

Obviously, the six collisions do not occur in any order. The first collison could be last or the last, first. They are ordered this way to supply a mixed-weapon, training progression.

Stop 1: The common stand-off, "showdown," or interview or ambush situation.
Stop 2: The common fingers, hand-on-hand or hand-on-wrist grabs.
Stop 3: The common forearm-to-forearm stop.
Stop 4: The common biceps-to-neck-line-to biceps stop.
Stop 5: The common bear hugs and/or clinches stops.
Stop 6: The common ground fight stop.

Stop 6 Stopping Point 1: The Showdown / Stand-Off

You walk up to trouble and trouble walks up to you. This common confrontation begins or stalls at the stand-off/showdown position, where two people are typically standing within a lunge and reach of each other and there is an exchange of words, perhaps at some higher volume and intensity. Usually, not always. Con men approach with a smile or a trick. It may not be obviously hostile in the beginning.

Interview or ambush? This event could be a "set-up," a disagreement, a confrontation over anything, or the tricky beginnings of a crime such as mugging, robbery, kidnap, or an arrest and detainment in progress. Maybe, a terror attack!

In the case of firearms, *Stop 1* parameters could conceivably be as far away as sniper range.

Stop 1 is the stand-off, "showdown" of first contacts, tricks, arguments and confrontations.

Stop 6 Stopping Point 2: Caught Red-Handed! Common Hands-on Stops

Ever watch the TV news and see a fight in the Taiwan parliament or someplace like the West Palm Beach City Council meeting? There you will often see people in these hand-on-hand, stop situations, like a palm-to-palm stop, of some sort. It is really not an uncommon stop or catch. Fingers entwine. Hands grab hands and wrist.

If you charge someone to grab them, push them, pull them or grapple with them, or even arrest them, you do so with hands up. They often respond with their hands and arms up too, which is why this catch is a common occurrence. This contact could be:

- fingers in fingers.
- hands on hands.
- hand grabs on stick, knife, gun and/ or carry sites.
- Stop 2 disarming.
- hands on fingers.
- hands on wrists.
- weapon retention.

Stop 2 will be covered in book 2 of the *Training Mission* series books.

Stop 2 is moving in! It's about the grabbing of the fingers, hands and wrists of the opposition.

Stop 6 Stopping Point 3: The Forearm Collision Stop

This is a stop/collision when opponents crash at the forearms. Those familiar with the martial arts might call this "Wing Chun Country," as this splinter of kung fu concerns itself with forearm collisions. This contact could be:
- any combination of the insides/outsides of forearms hitting each other.
- one person grabs the other's forearms.
- both people have a hold on each other's forearm.
- arms could be high or low.
- hand, stick, knife, gun.
- a combination of above.

Stop 3 will be covered in book 3 of the *Training Mission* series books.

Stop 3 is all about forearm collisions.

Stop 6 Stopping Point 4: The Biceps-Neck-Biceps Line Stop

Another common stopping point is the left biceps-shoulder-neck line-to right biceps. The arms are usually outstretched. A very common struggle position, similar to the starting position of much Judo practice and competitions. This could be:
- hands on biceps.
- hands on shoulders.
- hands on throat/neck.
- arms wrapping arms.
- hand, stick, knife, gun as in any combination of the above.

Stop 4 will be covered in book 4 of the *Training Mission* series books.

Stop 6 Stopping Point 5: The Bear Hug

This collision is some form of a bear hug, in sports or "street fights." Arms are caught in, or are free, or one arm in, one arm out. Sport people will often think of the classic "boxing" sport clinch. All clinches are bear hugs but not all bear hugs are clinches. This contact could be:

- chest to chest.
- chest to back.
- chest to shoulder.
- hand, stick, knife, gun.
- leg tackles.
- combinations of the above.

Stop 5 will be covered in book 5 of the *Training Mission* book series.

Stop 6 Stopping Point 6: The Ground Collision

One of the common ways we hit the ground is, simply put, that we fall. That is reason enough to learn ground fighting. But we are knocked down, tackled and pulled down. This ground contact could be:

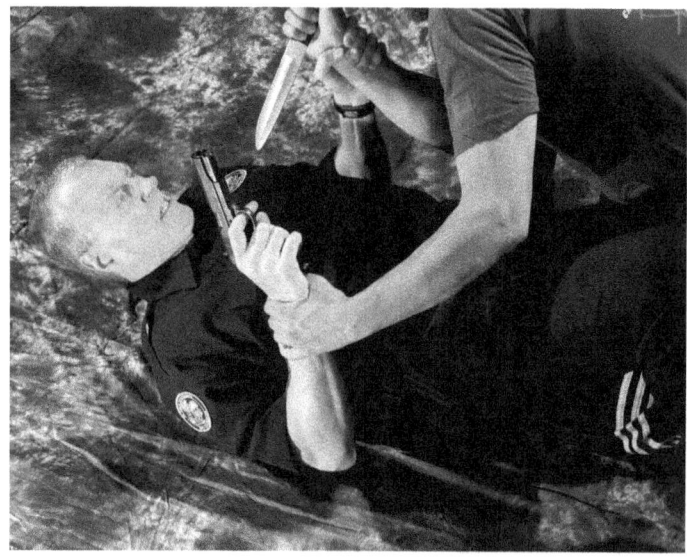

- you are on the bottom in some position.
- you are on the top in some position.
- you are on your right side in some position.
- you are on your left side in some position.
- hand, stick, knife, gun combinations of the above.

Stop 6 will be covered in book 6 of this *Training Mission* book series.

Stop 6 Summary
The Stop 6 concept is the experimental framework and laboratory for problem-solving. You insert things you learn from the *Force Necessary Hand, Stick, Knife* and *Gun* courses. The less you are trained, the more you will be stuck in these collisions. The more you learn and experiment, the better selections you can build to problem solve each predicament. With each stop you must try out:

- verbal skills.
- positioning and moving.
- striking.
- kicking.
- grappling.
- standing.
    * nothing around you.
    * pressed against a proverbial "wall."
- kneeling (maybe seated)
- ground.
    * topside.
    * bottom-side.
    * on right-side.
    * on left-side.
- unarmed vs.weapons.- weapons vs weapons.

The Stop 6 is the skeleton of these *Force Necessary* courses and *Training Mission* books. It structures methods in digestible, common sense order. With these you will never run out of workout, classroom or seminar topics. The deeper you go into the *Force Necessary* courses, which are each meant to be excellent, deep, realistic, "black belt," college-like programs, the more smart, training methods and options you will find.

In the unarmed course, the practitioner is sans weapons and cannot completely follow the stick, knife and gun *Stop 6* skeleton. It still emphasizes the vital, pre-fight information with many parallels.

The Force Necessary courses, and the Stop 6 concept offer a warehouse full of important unarmed and mixed weapon, training methods and topics, for personal practice, for schools, academies and seminars.

Stick Stop 6 summary

Stop 1: The common stand-off, "Showdown," or interview stance situation.
Stop 2: The common "hands-on" as in hand-on-hand or fingers, or on wrist.
Stop 3: The common forearm-to-forearm stop.
Stop 4: The common biceps-to-neck-line-to biceps stop.
Stop 5: The common bear hug or clinch.
Stop 6: The common ground fight.

Stop 1: Stand-off/Showdown.

Stop 2: Fingers and hands-on fingers, hands and wrists.

Stop 3: Forearm collision.

Stop 4: Biceps, shoulders, neck.

Stop 5: Bear hug/clinch.

Stop 6: Ground.

Chapter 3: Who, What, Where, When, How and Why Questions

"A problem well stated is half solved." - Charles Kettering

"Who, What, Where, When, How and Why." The phrase was first presented to me in the Army military police academy in the early 1970s. It was a checklist on how police officers should write a simple report. Answer those questions, big and small. But later I learned that a detective must further answer these questions, and a prosecutor must delve even deeper because you never know what weird little thing might become vitally important in a trial. These questions are also "journalism 101."

Then I learned I could apply all the questions to training objectives, and then...to all phases of life, really. Yes! I could certainly apply them to self defense, training and protection. And also to buying a house. Getting married. Raising kids. Even trying to get to safely use a neighborhood ATM. Even planning a military invasion like D-Day. Body-guarding the president? Answer these W and H questions as a framework.

I have used this "Ws and H" idea for about 30 years now as a spine and mainstay of my personal protection jobs and my training courses. I introduce these "Ws and the H" in Level One of all my training programs to set the stage for all subsequent levels. As the levels go on, I cover one question at a time.

Others have latched onto this "Ws and H" bandwagon too, certainly everyone I have instructed. Then some believers I haven't instructed personally, but have read my ideas or seen my films. Some were inspired by me. Others were inspired by me and just won't admit it. And some others found it by themselves. Still, few pound this important "Ws and H" drum as hard as I do. I didn't invent it. I just use the hell out of it. You should too.

Using it really takes about three to six rounds, or passes through the questions to really cover them well enough, because you realize you need to jump back to a previous W to answer the next W and so on. And, you certainly need the latest, unbiased, solid intel to evaluate your answers.

This intelligence usually comes from news, even gossip, media (non-fiction and fiction TV and movies), your experiences, or the experiences of others. Your brain processes this info and can give you "gut feelings/gifts of fear." (Much more on this topic later.)

With these survival and preparation questions, two variations are important to consider. First, all have dual, or duality issues. Second, all have the "big and the small," the macro and micro to ponder.

Duality

By duality I mean a "you" and a "him" duality. Or an "us" and "them" quality. The classic idea of dualism is really in several spiritual, religious, and philosophical doctrines. It's confusing sometimes. For example take the duality of the first "Who" question. You must answer "Who are you?" and "Who do you think you will be fighting?" You and him.

Big and Small: Macro and Micro

The second variation covers micro and macro answers. The answers can be big as in concepts or small as in very detailed and specific. An example of that? A "what" question. What will he do? He will rob you. "What" will you do? Try and stop him, for the macro, or big plan. Then move on down to smaller specifics such as if you move here, precisely what will he do next? He reaches into his pocket. What do you do? The micro.

Another example is when I ask you the big questions like, "Where do you think you might be mugged?" You might answer with something big like, "At the ATM." Good, big answer. But questions like the "When" question also has many little "Whens" to it, that are important to counter-tactics and survival. Like when does he step in too close to you? When does he actually pull out his gun? When does he actually turn to leave? When exactly do you fight back? Or run away?

All fighting is situational and positional. The big "When" question is the situational part. The little "When" questions are usually the small positional parts. A lot of self defense and fight training starts with the situational and then eventually concerns itself with positions. These answers, the small and little physical steps of the enemy are important when planning to fight or run for your life.

All these "Who, What, Where, When, How and Why" questions in this book and the *Training Missions* are about acquiring knowledge, developing awareness and problem-solving for both crime and war. I am going to offer some examples here to kick off your list.

Some "Who" Samples?

Who are you...really? Your job? Your physical condition? Your legal definition?
Who do you really think you'll be fighting?
Who are you to carry a weapon? Or not carry?
Who will judge your actions?
Who will take revenge?
Who will teach you to fight?
Who are nearby witnesses?
Who will come to help you?
Continue asking who questions...

Some "What" Samples?

"What happens next" continuum. The biggest question.
- do you fight and get arrested?
- get sued?
- get home safe, or back on base safe?

What will he do, what will you do?
What weapon will you have? His weapon?
What training course should you take?
Continue asking what and what-if questions...

Some "Where" Samples?
   Where, in your life and travels, will you have problems?
   Where? Being at the wrong (or right) place at the wrong (or right) time!
   Where will he hide to ambush you?
   Where can you run, hide or fight?
   Where? Suspicious places. Dangerous places?
   Continue to ask where questions...

Some "When" Samples?
   When are the dangerous times to be out and about?
   When are you most vulnerable? Least vulnerable?
   When can you get help?
   When exactly can you use a weapon?
   Continue asking the when questions...

Some "How" Samples?
   How will you react?
   How will he react?
   How will you survive?
   How do you train for an ambush?
   How much - "gas" (lasting power) and - "TNT" (explosive power) do you have?
   Continue asking the how questions...

Some "Why" Samples?
   Why go there?
   Why are you near there?
   Why are you staying?
   Why is he fighting?
   Why use a weapon? Why not use one?
   Continue to ask the why questions...

Continue to delve more deeply into each question.

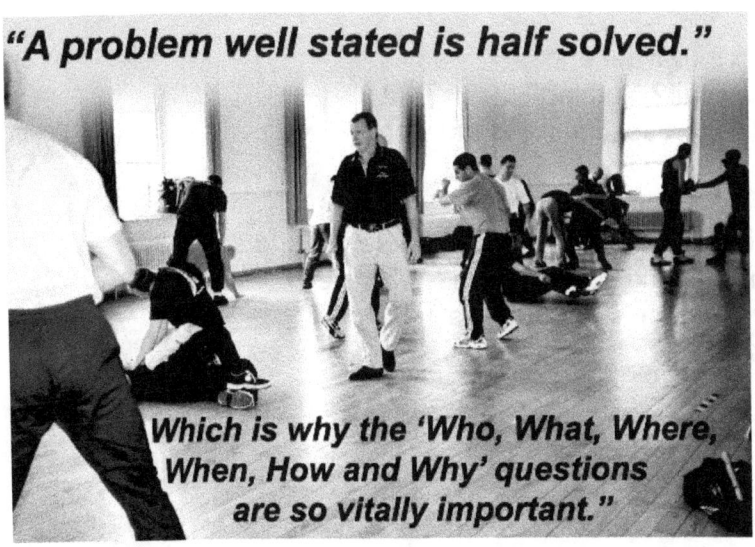

"A problem well stated is half solved."

Which is why the 'Who, What, Where, When, How and Why' questions are so vitally important."

Photo taken at the old SAS, Duke of York's Barracks, England.

That was a generic look at the W and H questions, now let's break it down to impact weapon study.

Who Stick?
What Stick?
Where Stick?
When Stick?
How Stick?
Why Stick?

Who Stick?
Who are you to be walking around carrying an impact weapon?
Who do you think you will be fighting?
Who will teach you how to use one? And not just a martial artsy version.
Who will judge you for using an impact weapon?
Continue to develop who questions and answers.

What Stick?
What stick will you carry?
What will happen that you think you will need a stick?
What stick course will you take?
Continue to develop what questions and answers.

Where Stick?
Where on your body will you carry a stick?
Where do you think you will need one?
Continue to develop where questions and answers.

When Stick?
When will you need a stick?
When will it be legal?
Continue to develop when questions and answers.

How Stick?
How will you use your stick? To what end?
How will this end? Jail? Lawsuit? Nothing?
Continue to develop the "How" questions and answers.

Why Stick?
Why do you think a stick is a good weapon for self defense and survival?
A trouble spot or situation. Why did you go there? Why are you still there?
Continue to develop why questions and answers.

More "Ws and H" Advice

This is a book about impact weapon combatives, not a law book, or even a "law of the jungle" book. It still behooves me to remind every reader about the legal concepts of selection, carry and use of any sort of impact weapon. You MUST check with the local laws of where you live and where you are going. They vary. Some police agencies forbid the carry and use of baton and night sticks. In the military you have your rules of engagement. These are indeed who, what, where, when and how issues.

I worry about the "who, what, where, when, how and why" questions. In my courses and should be in your courses too, part of the "Who Question is "who do we fight?" Well, we fight three "enemies."

> 1: Your "drunk uncle"
> 2: Criminals
> 3: Enemy soldiers

1: Who? Drunk Uncles: "Drunk uncle" is a metaphor that means all your relatives, near and dear, near and far. Kin folk or those close enough to be. It is very common in life to fight people that you do not wish to really hurt. Like your drunk buddy or uncle/relative. In police work we are also expected to fight but not really hurt people unless things get really "out-of-hand" and the situation escalates. But in person-to-person, poke your buddy's eye out, bite off his ear, hammer-fist his throat or neck, smash his face, break bones, shatter his knee, and then see what happens to you. Usually, often, jail and lawsuits. Lots of money and problems. There is a whole lot of domestic violence out there, and violence on, and from, "who you know" is a big problem. (Remember, there are many intricacies in the complex laws of family violence, lest of all assaults and self defense.)

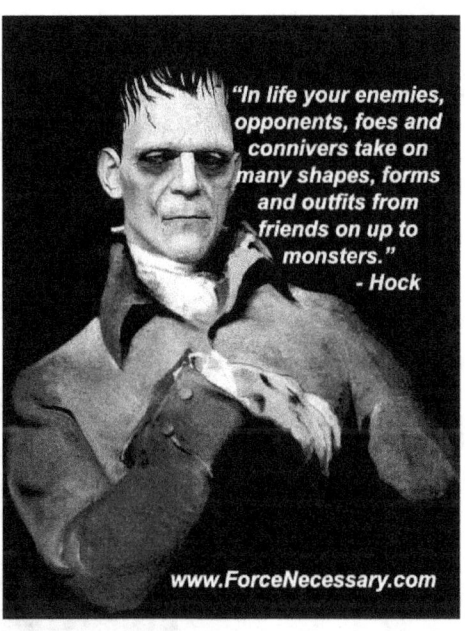

"In life your enemies, opponents, foes and connivers take on many shapes, forms and outfits from friends on up to monsters."
- Hock

www.ForceNecessary.com

2: Who? Criminals: Essentially speaking, a stranger, (or for that matter even a friend, uncle or not, officially becomes a criminal when they assault you. You could just lump your uncle into this category once in a while too. But, what crime is being committed? Who, what where, when, how and why? The level of crime, the exact situation takes the exact temperature of your hot, lukewarm or cold response. Crime by the way often starts out with a trick ambush, which is a deep dive study also into the "what, where, when and "how" questions.

3: Who? Enemy soldiers: We know what those are. We usually like to kill them from as far away as possible, but often can't do that either. Consider the military "rules of engagement."

Civil law, criminal law and the Geneva Convention, as well as human ethics – look at fighting these three "bad guys" categories differently. Our responses and solutions confronting said "uncles, criminals and enemy soldiers" are very situational and may be:

>**Surrender.**
>**Bargain (talk, show weapon, etc.).**
>**Escape (orderly retreat – you leave or he leaves).**
>**Hurt, on up to maim.**
>**Kill.**
>**Detain, arrest and-or take prisoner.**

Of course, not necessarily in that order. All are worth exploring in training through the "who, what, where, when, how and why" questions. All have happened and will happen. I make it a point to cover all of the above in the Force Necessary courses and the *Training Mission* books.

Since we are Force Necessary and not Force Un-necessary, I do not teach sports or arts. I have done sports and arts for decades. I investigate sports and arts. I only borrow and raid from sports and arts for practical applications to solve these "uncles, criminals and enemy soldier" problems. Sports and arts are great laboratories, but it takes constant vigilance to know where to draw the line between art-sports and survival.

Pre-Assault? What about Pre-Crime?

Another issue in "Ws and H' question. When do we fight back? There has been much ado these last years in training/seminar circuit about pre-fight indicators. Instructors present a list that has actually been around since the 1970s. So new? No. Just new to new people, that is. Through those early years the list rarely filtered down into the local "kuraty" clubs, so to speak, so its arrival decades later, was big and big business for some. It is shocking to me that so many martial veterans were unaware of these set-ups.

It seems that most pre-fight indicator lists, and their courses, however have been mostly about "boys in bars fighting." Not about criminals and crime. NOT a pre-crime confrontation list! The pre-assault advice covered is usually what an angry person does just before he or she hits you. Which is a crime, but not always a premeditated criminal plan-ambush. Not that these emotional "sucker punches" aren't important too, and criminals about to attack you also have biological symptoms too. So for the record, we list are the classic tips.

What are Some Tip-Offs He May Attack You? This info was first taught to me in the 1970s at military and Texas police academies. I've collected it all, adding some, and the list is in my teaching outlines since the 1990s and in *Fightin' Words* book. My *Training Mission One* book is all about hand, stick, knife and gun Stop 1 "collisions," that is, all the things that happen before physical contact from sniper range to stand-offs. Since you are not reading those now, you are reading this, here are some of those trouble-tips.

Now, I do not want you to over-emphasize this information as some kind of cure. Just read over the list and keep them in mind. The list was created and repeated here because these tips/events have happened. I have seen many of them when dealing with people for decades in this upset and angry, drugged or drunk "people business" called police work.

When a person becomes stressed, angry and aggressive, his or her body might react, not always, but sometimes it demonstrates some changes. Here are some of these changes that research, history and experience may induce a sudden attack/leap upon you. Many people suggest that in a real fight situation, a person has no time to read these clues. Sometimes, yes, I agree. But, this is not always true. Sometimes there are confrontations and people do have the time to see these tip-offs. Every professional and every citizen needs to read this list and at least become aware of these points.

Obviously the clues vary from situation to situation and person to person. But, better to know the ones on the list, than not, or to ignore they even exist. I have seen them unfold myself on police calls and making arrests.

His eyes bulge.
He has that 1,000 yard stare.
He suddenly seems to ignore you.
He squints.
He assesses your body parts and gear as potential targets.
His mouth becomes dry, creating odd lip and jaw movements.
His teeth clench.
His voice changes.
He actually, clearly voices violent intentions.
His words become spastic and distracted.
He twitches.
His nostrils flare.
His breathing increases.
He takes one big sudden breath.
His face color changes, maybe reddens or pales.
His veins bulge.
His chin tightens, or drops.
His neck tightens.
His jaw juts (dumb but he still does it).
He babbles as though his thoughts are not guiding his voice.
He doesn't babble and actually vocalizes his plans of attack.
He actually tells you his plans! "Why I'm gonna…"

His arms swing, maybe with body turns (a big deal and easy cover for a sucker attack).
His fingers and fists clench (blood leaving those extremities).
His fingers drum surface tops.
His hands shake.
He extends a hand to shake yours. Could be a trick.
Hands go to weapon, carry sites on the body (previously listed).
He turns away (critical sucker punch set-up).
His hands and arms travel to near obvious pre-fight postures and positions. He positions his hands high on his chest, neck, chin or head. Raises up to seemingly innocent, high positions as in a fake head scratch, like a yawn or a stretch.

He strikes a pre-fight posture, such as a boxer.
He raises from a seated position.
He tries to wander.
He bends slightly at the knees. (A sporty-like body crouch is never a good sign. I want to say in my experience that I have found one of the biggest tip-offs to trouble is a crouch! Bending at the knees. When the other person crouches. This is a spring board to athleticism. Not only might they attack you, or run off, but in the mixed weapon world we live in, people have a tendency to crouch and draw knives and guns.)

He gets too close.
His body blades away from you.
He suddenly takes off his shirt, jacket or watch.
He "expands" his chest.
Heel and toe tapping.
Positioning near potential improvised weapons.
Shirt lift about his belt line (this is NEVER a good thing).
Keep adding to this list.

Pre-crime. But, what of pre-crime indicators? Planned criminals can display none of the signs. They can smile, act and approach with a trick, gimmick or question. I am not sure that the average Joe and Joan grasp the fact that the thrilling, pre-fight indicator list can be quite different than the pre-crime indicator list. Oh, and I can hear the snoring already beginning because this now reads like…"crime prevention." BORING! Huh? Crime prevention is often cluttered with "locking your doors," and "putting up outdoor lights," and…and…still awake? Still reading?

How does one…pre-crime? How do you detect an ambush crime? Pre-crime studies are different than pre-fight studies. And I believe that while many virgin schools and virgin seminar attendees are so happy to hear about all the "fist clenching" and "1,000 yard stares," that the presenter and attendees miss the crime prevention aspects.

Collecting criminal intelligence in general and in your area is important. Stopping rapes, robberies, abductions/kidnappings, home invasions and murders. Who, what, where, when, how and why do you get ambushed into a crime? Sometimes there's a little overlap between the two categories, sure. But pre-crime is different and diverse. For example, there are usually little if any pre-fight indicators in a criminal ambush. Many criminals just ambush you from behind. The element of surprise has defeated the greatest militaries of the world and it can defeat you too.

What can we do to make pre-crime sexy again? It's hard. Publishers use to create a fair amount of crime prevention books years ago. They were quickly rendered onto the dollar sale table. No sales? No more books.

People do somewhat remember *The Gift of Fear.* Why? The stories, that's why. Years ago, Gavin Debecker wrote that entertaining book, *The Gift of Fear.* First editions really promoted an ESP-ish, Spidey-Sense as the gift. Neuroscience developments in the 2000s proved otherwise – that it wasn't magic, rather we react from learned behavior. Your "gut" instinct is almost completely a trained mind from vast sources. The gift stories were thrilling (psychology has already proven that stories and "war-stories" are the best, longer-lasting teacher). But take out the cool stories? And what's left, the skeleton of advice? Strip out the tales and you have a BORING crime prevention hand-out from your local police department. "Lock your doors." "Put up lights." "Watch out for strangers." "Watch out for dark places." Etc. Yawn.

The routine crime prevention pamphlet can leave something to be desired. It usually lacks a certain first-person, in-the-moment advice from...stories. Whereas watching a news story about an unlocked door, and a sobbing crime victim, is a better teacher than a McGruff pamphlet.

Geography, plus architecture, plus criminal mind. For one example of a study area for pre-crime in the "where" category, I wrote about this in my book *Fightin' Words.* I worked a rape once by a bus stop. In the daytime, this ¾ enclosed bus stop looked normal and safe. A curved sidewalk ran behind the little clear, plastic edifice. In the middle of the walkway, beside the curve was a small grassy area, then tall fences of an apartment complex. This area had a gigantic bush-looking tree next to the sidewalk. Looks safe and normal. In the daylight. But at night? It was a trap. Poorly lit. A college girl walked by and was snatched by a thug from behind this bush. When called out to the case, I saw this scene at night and could see what a trap it was, from a criminal mind perspective. Daytime? No. Nighttime, yes.

An equation for trouble. Who, what, when, where, how and why? These questions can be investigated with good intel, research, experience, and an adequate mind, to predict crime scenes. With the "who, what, where, when, how and why" questions.

Who are you as a victim? Study victimology.
What crime could occur?
Where are you most or partially vulnerable to crime?
When are you most or partially vulnerable to crime?
How will the criminal approach?
Why are you there? Why are you still there?
This is just the beginning of the exam...

Hey, let's make crime prevention interesting again! I mean, doesn't "Pre-Crime" sound cooler than "Crime Prevention?" We can do this. Keep your "scene" just a "scene" and not a crime scene...baby.

Chapter 4: Selecting Your Impact Weapon
Weapon Selection 1: Mission Needs

Define your mission. Using the latest intelligence information, predict your probable conflict needs. Sometimes a long flashlight is all that's required, sometimes the smallest expandable baton. Sometimes the biggest riot stick. All encounters are situational. Prepare for the mission.

Weapon Selection 2: Your Body Size, Strength, Shape and Arm Length

These factors will dictate the type of weapon you select and your skill at manipulating it. At best, your impact weapon needs to be long enough to be an effective crowbar for leverage with one or two hands, yet short enough for you to execute all your single-handed striking tactics, as demonstrated later. Another classic test is to hold a stick in your hand at arms length. If you can circle the stick in a full rotation at the wrist and, if the stick can touch your head in its turn, it will be too long for many one-hand combatives techniques. But, if you need and are assigned a weapon that is long like a riot stick, you will probably be using a two-handed grip anyway.

A common walking cane or the martial arts stick are impact weapons, as is any similar object of length.

This Frankfurt, Germany police "stick" was given to me after teaching their area SWAT teams. As you can see, it is very short and made of a very hard rubber.

Expandable batons come in a variety of sizes. These three are the typical sizes from 12 inches to 28 inches. Longer riot-based sticks also exist.

Various things that are impact weapons.

How about the ubiquitous axe handle?

Any sort of long solid "thing" that can be held with one or two hands.

What Stick Should I Select?

What is your mission? Review the Ws and H questions. The stick you need is selected by your "overall mission." Do you need a long, riot stick? A short stick, or short expandable baton for your pocket? Long ones? Don't get, carry and do something that doesn't relate to your who, what, when, where, how and why.

Impact weapon combatives concerns itself largely with:
   1: Stick versus unarmed.
   2: Stick versus stick.
   3: Stick versus knife.
   4: Stick versus gun threats.

Can You Carry and Use a Stick?

In some jurisdictions, in some cities, counties, states and countries around the world? Yes. You must check the jurisdictions where you live and where you travel. Check laws on "brandishing a weapon." The fighting use of a stick is based your local legal system, sometimes tricky, and on self defense and common sense.

The Three Stick Carry Sites

Fixed sticks, even collapsed batons can be difficult to carry, so know all about the three weapon carry sites for yourself and your enemy. They are:

   1: Primary carry sites; Think easy weapon quick draws, like the belt line and armpits. With collapsed batons, pockets are also very quick draws.
   2: Secondary carry sites: Think back-up sites. It takes some "digging" to get them.
   3: Tertiary carry sites: Think lunge and reach "off-the body."

Chapter 5: Drawing the Impact Weapon

In order to use the impact weapon, you have to draw it out from a carry site!

Many police officers, military personnel and security officers carry batons. Many are issued "straight," simple sticks that are carried on their uniform belts in partial leather holsters, or hung by rings on straps. If the officer carries a pistol on his strong side, he will carry his baton on his weak side. Many officers and guards around the world do not carry firearms and are issued impact weapons. In this case, these non-firearm carrying personnel will carry their sticks on their strong side. This right or left-handed draw of a right or left belt carry-site makes for a mix and match that creates a mathematical puzzle for subsequent tactics and strategies.

Two Moral, Ethical and Legal Reasons to Draw Your Weapons
    Reason to Draw 1: Stop violence before it happens.
    Reason to Draw 2: Stop violence while it is happening.

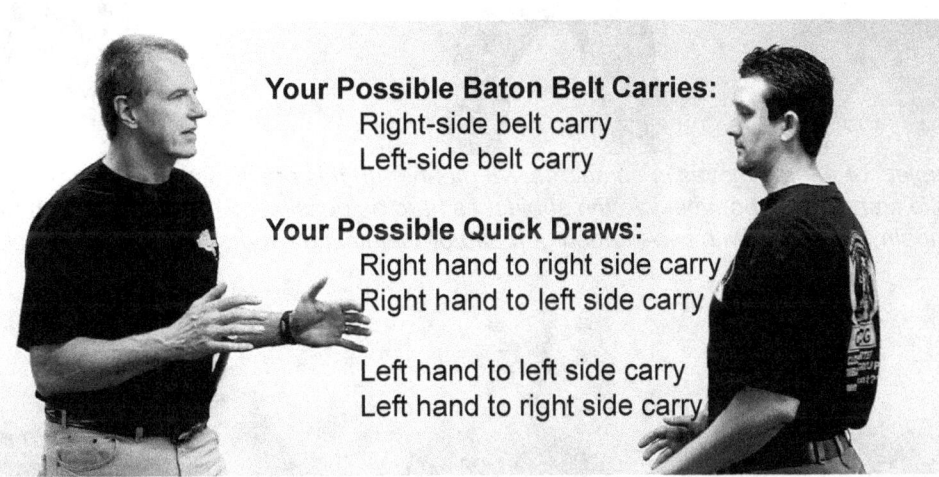

Your Possible Baton Belt Carries:
    Right-side belt carry
    Left-side belt carry

Your Possible Quick Draws:
    Right hand to right side carry
    Right hand to left side carry

    Left hand to left side carry
    Left hand to right side carry

Belt - Left hand to left side draw.

Belt - Right hand to left side draw.

Belt - Right hand to right side draw.

Belt - Left hand to right side draw.

Your "Straight Stick" Police Quick Draw Workout

The following straight stick, quick draw steps are taught to many law enforcement officers in the United States and around the world. It starts with a left hand draw from a left hand carry site since most officers are right-handed and have a pistol on their right hip. Of course, practice both sides.

Step 1: Check your required uniforms. Select your probable carry site.
Step 2: Practice drawing from that carry site.
Step 3: Practice drawing with a person in front of you. Learn position and the range between you.
Step 4: Then practice from seated, kneeling and ground positions.

The subject or enemy moves in an aggressive manner. In this practice police scenario, you draw and strike first with the pommel into the subject as hard as possible. Grab the stick with two hands. Hit again, if needed, with two-handed grip and/or maintain a two-handed grip, if you choose.

Or, hand-off to a single-hand grip and strike. Most police agencies require a low, thigh nerve strike at this point, even though this thigh strike is almost always ineffective in actual situations. In this photo series, I am striking high.

Drawing the Expandable/Telescopic Batons and "Short Sticks" from a Belt Carry

Whether civilian clothes or uniform. you need a holster/sheath for your weapon. This belt line draw, expandable baton draw is the same as the fixed, long stick draw seen earlier, but once out, we must still open the baton.

A closed telescopic might be considered a short stick, akin to those popular in Korean martial arts.

Belt Draw 1: Left side carry, left hand grabs and draws weapon.
Belt Draw 2: Left side carry, right hand crossed over and draws weapon.
Belt Draw 3: Right side carry, right hand grabs and draws weapon.
Belt Draw 4: Right side carry, left hand crossed over and draws weapon.

Pull apart expansion by hand into a locked position for larger, heavier models: A sample. Some of these have release buttons!

Expandable/Telescopic Baton - The Swing Momentum Opening

Here is a momentum quick draw of an expandable baton. In this one, an operator needs whipping strength and enough space for the telescopic action to swing open and lock into place.

Throw the tip back! Draw and extend the expandable baton. The momentum expansion: One sample.

Expandable Baton - The Jerking Momentum Opening

Here is a momentum quick draw of an expandable baton that does not require a swinging motion and a lot of extra space. It requires a snapping, jerking motion of the arm that propels the baton out and into position, whether up or down. This move is especially handy in ground fight situations.

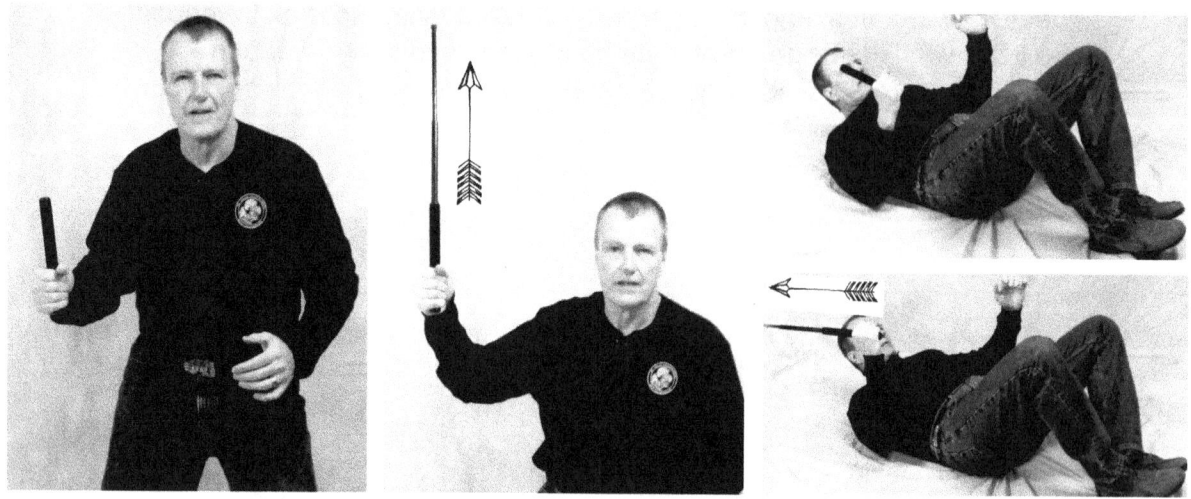

Your Expandable Baton Straight Line Snap, Quick Draw Workout

Step 1: Check your required uniforms. Select your probable carry site.
Step 2: Practice drawing from that carry site.
Step 3: Then practice drawing with a person in front of you.
Step 4: Then practice from kneeling, seated and ground positions.

Walking and Running Quick Draws with both Straight Sticks and Expandable Batons
The trainee starts from a distance. He approaches the trainer at a walking pace or a running pace. Or is being chased! The trainer draws, and the trainee must respond with a draw while in motion.

Walk or run and draw on any realistic cue.

Unarmed? No. I've got a stick up my sleeve. Look for those telltale, curved fingers holding something up.

The Stop 6 Stress Quick Draw Practice

These are the six common stopping points in a two-person encounter or an arrest situation. Even if you are inside a hostile group, slivers of that group fight often contain these one-on-one encounters. The less you are trained, the more you will likely become stuck in any of these positions. These are not about ranges, these are about situations and situational fighting. Your training is not complete without training in the Stop 6. To review, the Stop 6 situations are:

Stop 1: The common stand-off, The "Showdown," or interview situation
Stop 2: The common Hands-on. Hand-on-hand, hand on arm. Fingers may weave.
Stop 3: The common Forearm-on-Forearm stop
Stop 4: The common Biceps-to-Neck-to-Biceps stop
Stop 5: The common Bear Hug or clinch
Stop 6: The common Ground Fight

Practice your impact weapon quick draw in all six of these situations. At this stage of the progression, simply practice the simple weapon draw. Other extenuating factors will be introduced through the Stop 6 as this program continues.

Practice 1: Draw a straight stick from your carry site through all the stops.
Practice 2: Draw an expandable baton, and open it through all the stops.
Practice 3: The Quick Draw Combat Scenarios.

Draw Combat Scenario 1: Draw and hit attacking hands and feet of an assailant
Draw Combat Scenario 2: The Shield. Knocked down, scramble, floor walk with shoulders and pull weapon.

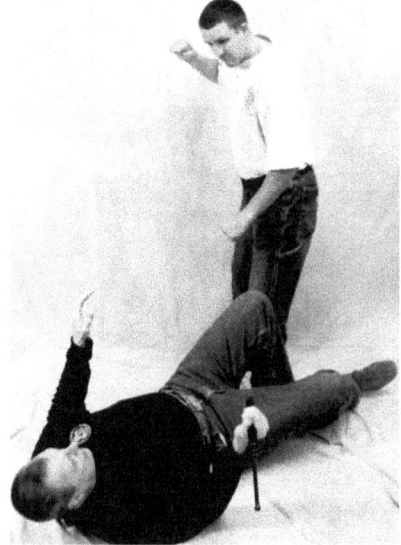

You are knocked down. Draw and swing your stick as a shield so you can get back up.

Practice drawing under the stress of an attacker. One trick is to draw from a rear back hip position.

Sample of Stop 4 Combat Stress Quick Draw Scenario 1

You are being choked. You start to draw your weapon, but the choking and pinching pressure increases on your windpipe. You cannot risk a few seconds hitting the attacker with your stick. You must seek a fast release from the grip.

He chokes. *You draw.* *But the choke grip increases.*

*No time for strikes. You collapse the arms, free the choke
and begin striking your way out of the fight.*

Counters the Choke Scenario
- 1: The enemy grabs your neck and squeezes.
- 2: You feel you have a few seconds to take some maximum measures.
- 3: You draw your baton and open it. Roll it over the bends in his arms.
- 4: In a two-hand grip ram the arms down, clearing the choke.
- 5: You then ram his face with a savage impact.
- 6: Keep punching and striking forward until the enemy is back and off.
- 7: Exercise this with you on the bottom-side ground. Thrust the enemy off.

The Shield Stress Quick Draw Combat Scenario 2

This is a sample scenario for when you are knocked down and about to be jumped. You quick draw your impact weapon and create a shield as you get back to your feet.

Barnhart is punched down.

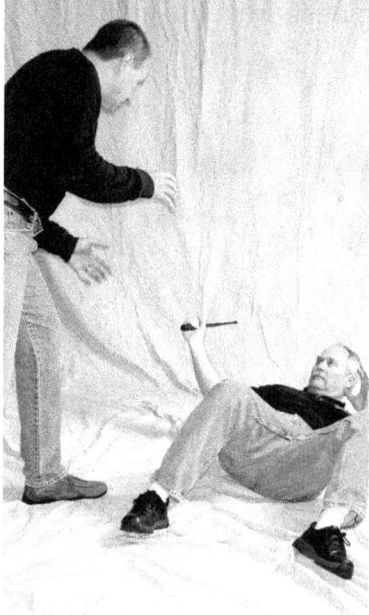
Barnhart draws his impact weapon.

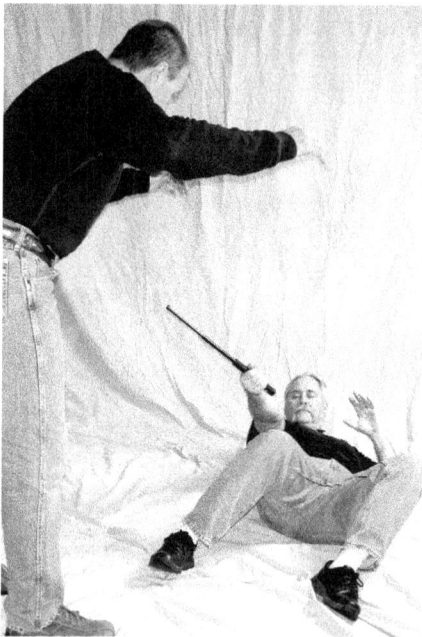
He swings as soon as possible.

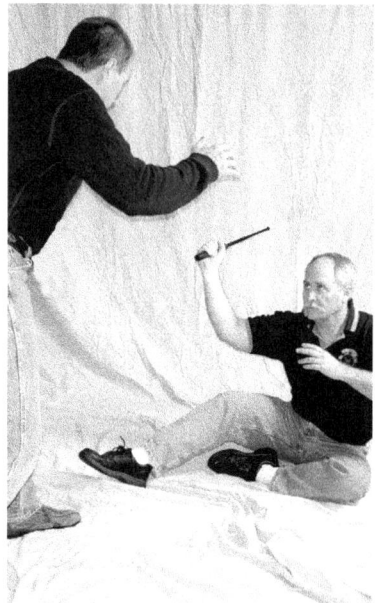

The Shield
1: The enemy strikes you down.
2: You gain some space with a shoulder walk and/or grounded thrust kicks.
3: You draw your baton.
4: You immediately start swinging at the enemy.
5: This creates a shield, keeping him away from you as you get up.

Remember When You Need to Draw

Remember the stances, carry sites, gadgets and tricks of the weapon quick draw that were shown in the prior chapters. The best quick-draw? Getting the stick out just before you really need it. But review the prior section on brandishing a weapon. The weapon pull should be justifiably explained.

Factor 1) Use of force, rules of engagement and local weapons laws.

Factor 2) Fixed sticks of varying lengths.

Factor 3) Expandable batons of varying lengths.

Factor 4) Belt carries with breakaway holsters or ring straps.

Factor 5) Momentum openings for expandable batons.

Factor 6) Manual pull apart openings for expandable batons.

Factor 7) Spring release openings for expandable batons vary.

Factor 8) Clothing that carries or conceals, regardless of the weather.

Factor 9) Length of operator's arms and torsos.

Factor 10) Space available to open and/or draw these weapons.

Factor 11) Drawing while standing, moving, seated and grounded.

A Training Tip

If you are coming to a stick class or stick seminar, you should also bring a belt. You need to have a realistic carry system with you, so you can practice a realistic draw. These sticks don't just appear like magic in your hand.

CHAPTER 6: THE IMPACT WEAPON GRIPS

This *Force Necessary: Stick* course covers all the single-handed and double-handed grips on all impact weapons. Grips will vary depending upon the situation. So, once drawn, what's your grip?

Grip 1a) The standard, popular one-hand grip, saber-style.

Grip 1b) The single-handed, reverse grip.

Grip 2) The center grip, a one-handed hold in the middle.

Grip 3) The common sword or baseball bat grip.

Grip 4a) The rifle grip, one one palm up, one palm down.

Grip 4b) The stick grip with grip both palms down.

Issues with The Single Hand Grip. How much of a handle should you leave exposed as a grappling and striking tool protruding from the bottom of your hand? Many stick systems use the following method as a measuring tip for choking up. They wrap the support hand on the end and place the actual gripping hand above that. This leaves a few inches of stick exposed at the bottom.

Other stick martial arts systems have little or none of the handle exposed out of the bottom of the hand. These are stick-versus-stick martial arts and thus, expect faster, hard core impact strikes as their main tactic, such as to the limbs and the head. Most tactical stick courses never worry about such a thing. They will not be fighting a stick fighter. These practitioners consider any choking up on the stick, the same as baseball players choking up on a baseball bat. They state choking up takes away from their full-power, impact power swing. The other main reason these stick martial artists do not choke up on their stick is to prevent the ubiquitous Stick Snake Disarm. If the handle is exposed? In close quarters some may be able to execute this common disarm. This disarm is so common that totally untrained people have done it by accident! (The unenlightened and inexperienced think the snake disarm is way too technical and impossible to do. And then some crack cocaine addict does it to your baton or flashlight. Surprise! I am still collecting snake disarm stories, usually from untrained people). But, this handle-less fighter, fearing the snake disarm, loses many close quarter advantages such as hooking, grappling and striking with the sides and end of handle.

The tactical military, police or citizen defender will most likely not be involved in a 28-inch stick versus another 28-inch stick dueling match, as so many stick martial artists train for. Mixed weapons and empty hand opponents are involved. Grappling will be involved. And, an exposed handle is a great tool for grappling, handle striking and extreme close quarter work regardless of the enemy's weapon.

To choke up or not to choke up? One must also undergo a scientific study of real world, striking force. For example, how many foot pounds of pressure does it take to crack a skull? Criminal forensics tell us that it varies. It varies with the individuality of each skull. Thickness does vary. Many will argue that a skull is more easily fractured from the sides than at the top. So target location counts too.

One forensics source claims that "a force of 73 Newtons is enough to cause a simple fracture, this force is the equivalent of walking into something solid. An unrestrained adult fall from standing has been shown to produce a minimal force of 873 Newtowns which is more than enough to produce a skull fracture." Another source says "it takes about 15 PSI to crack the skull, while it takes about 7 PSI to crack a rib." This is not a science book about Newtons or about PSIs. Most experts say it is debatable and within some reason, it all varies.

Let's stick within reason. Let's pick some easily calculated, estimated speeds of 80 and 100 miles per hour for this essay. Many agree that stick swingers actually swing faster than these speeds. Assume you swing a stick with no handle exposed at 100 miles per hour. Then you choke up on the handle. You now swing a stick with a slightly choked-up grip with some some handle exposed at 80 miles per hour.

If you can crack a skull with a 40 MPH strike, or an arm bone at a 30 MPH strike, it really doesn't matter if the handle was exposed or not, does it? 100 MPH? Or 80 MPH? Does it? You have plenty of speed and power to spare, if you choke up just a bit to expose the handle, now you have a whole host of other tactics to use with that exposed handle.

That covers the discussion of impact. What about the need for speed in fighting set-ups, faking and overcoming blocks, etc. One might argue that 20 more miles per hour, 80 to 100, extra speed is needed to enter and swing at the enemy. But the difference between 80 to100 MPH? Really? Whom do you expect to be fighting? These higher-speed concerns are mostly very high-level, stick-versus-stick, block and strike issues, not tactical civilian, police or military issues.

The speeds attained with a slightly, choked-up/handle-exposed stick are more than sufficient to achieve reality, fight success. If you were dueling against a stick master with your stick, you might worry more about these martial artsy, masters' points. The duel. The myth of the duel. And, need I add that many of these old stick masters do quite well with choked-up grips on their sticks anyway, as is their system's preference.

A very common method to establish a common, proper grip. You grab at the bottom with your support hand. Place your main hand atop that support hand. Then let go of the support hand.

Stick versus stick fighting is mostly a martial arts, hobby endeavor with abstract benefits.

Who in our modern times will be fighting stick versus stick? Well, in some countries it can happen. And, at times modern people are fighting each other with stick-like, impact weapons, so a proper doctrine and practitioner should cover generic "stick versus stick" to a healthy, proper and wise extent.

Whether stick versus stick or not, an "exposed handle end (see left) can be used to hook, grab, and/or temporarily control an opponent's limbs and body.

As a regular practitioner and researcher of impact weapons for over three decades, I wholly recommend that you choke up on the handle of the stick...just a bit. Not too much.

CHAPTER 7: IMPACT WEAPON READY STANCES

You have drawn your impact weapon. It is out, and at the ready. The fight has not started yet. These following starting positions create command presence for the citizen, the police and even the riot squad. A stance has some advantages. It should get you ready for action. It might intimidate an opponent. Command presence. Remember though a fighting stance is about balance and power in motion. Not a still photograph.

Stick down and ready?

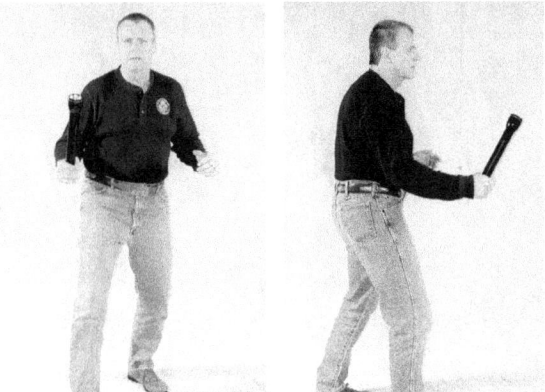
Position 1: One hand and weapon forward.

Position 2: One hand, weapon neutral.

Position 3: One hand, weapon back.

Position 4: One hand "police" baton back.

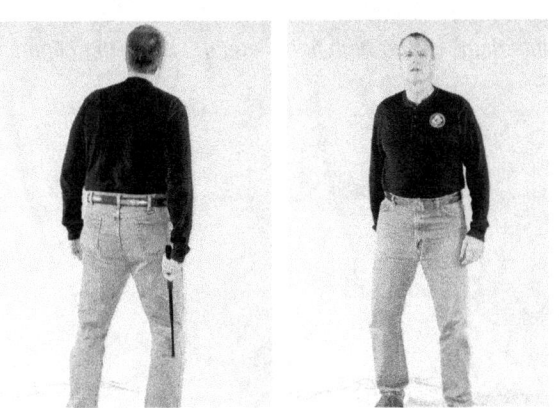
Position 5: One hand, concealed saber grip.

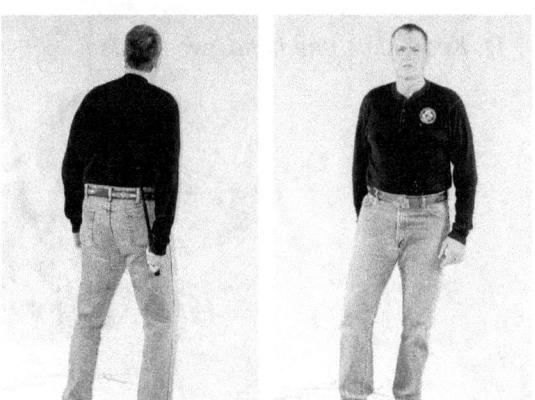
Position 6: One hand, concealed reverse grip.

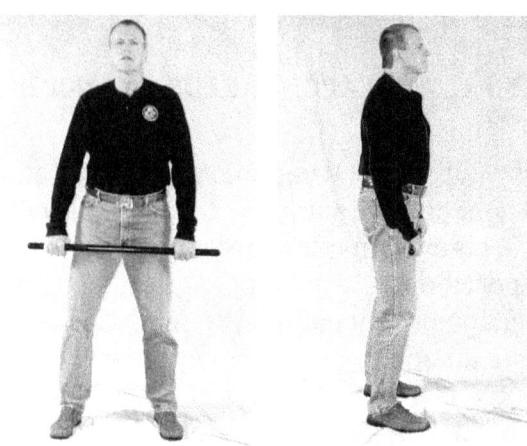

Position 7. Two hands, riot-ready thighs.

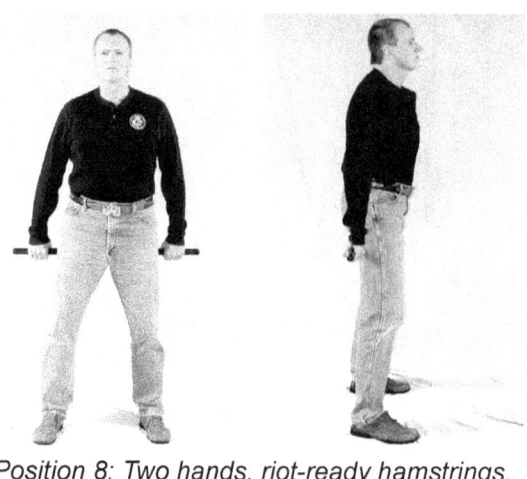

Position 8: Two hands, riot-ready hamstrings.

Position 9: Two hands, riot-ready port arms.

Position 10: Two hands, riot-ready thrust.

11: Knee-high, one hand, two hands   *Knee-high fights high*   *Knee-high fights equal*   *Knee-high fights low*

12, 13 and 14: Grounded, back and sides.

The Ready Position List

Keep in mind a stick down at your side *might* be a ready position, or not. A practitioner should know and understand all these 14 ready positions:

Ready Position 1: One Hand, Stick Forward
This is a favored position of many stick practitioners. The weapon is up front and protects the body.

Ready Position 2: One Hand, Stick Neutral
This weapon and body are somewhat in alignment, ready to commit to a more combative situation.

Ready Position 3: One Hand, Stick Back
This is often favored when the enemy has no weapon and the empty hand can be forward to block, strike and grab.

Ready Position 4: One Hand, The Police Ready
This is the favored, internationally taught fighting stance for police. It disregards many smarter positions and tactics in an effort to be uniform and simplistic.

Ready Position 5: One Hand, Stick Down and Concealed Saber Grip
For strategic reasons, the weapon is pulled and hidden from the enemy, with a saber grip. The shaft is hidden behind the leg. This has been a riot squad command to conceal the weapon and create a calmer image to the public.

Ready Position 6: One Hand, Concealed Reverse Grip
For strategic reasons, the weapon is pulled and hidden from the enemy, with a reverse grip. The shaft is hidden behind the torso. This has been a riot squad command to conceal the weapon and create a calmer image to the public.

Ready Position 7: Two Hand, Riot-Ready Thighs
The weapon is held across the thighs, visible to all. This is also a common riot squad command that is less aggressive to the public.

Ready Position 8: Two Hand, Riot-Ready Hamstrings
The weapon is held across the hamstrings, partially visible to all. This is also a common riot squad command that is less aggressive to the public.

Ready Position 9: Two Hand, Riot Ready Port Arms
The weapon is across the chest, present to an enemy or the public and somewhat aggressive. This is also a common riot squad command that is somewhat aggressive to the public.

Ready Position 10: Two Hand, Riot-Ready Thrust
   This is an aggressive position. This is also a common riot squad command that is aggressive to the public. The next step is forward action.

Ready Position 11: One Hand, Knee-High
   Be ready to fight knee high. You will fight the enemy in three height levels:
      Direction 1: above you as in a standing enemy
      Direction 2: equal to your knee-high height
      Direction 3: below you. This is your top-side ground fight position.

Ready Position 12: Two Hand, Knee-High
   Be ready to fight knee high. You will fight the enemy in three heights, levels:
      Direction 1: above you as in a standing enemy
      Direction 2: equal to your knee-high height
      Direction 3: below you. This is your top-side ground fight position.

Ready Position 13: One Hand, Grounded

   Be ready to fight from on your back and sides while on the ground in a single-hand grip.

Ready Position 14: Two Hand, Grounded

   Be ready to fight from on your back and sides while on the ground in a two-hand grip.

Summary, The Ready Position List
   Ready or not ready? Stick down at your side.
   Ready Position 1:   One Hand, Impact Weapon Forward
   Ready Position 2:   One Hand, Impact Weapon Neutral
   Ready Position 3:   One Hand, Impact Weapon Back
   Ready Position 4:   One Hand, Police Ready
   Ready Position 5:   One Hand, Concealed Saber Grip
   Ready Position 6:   One Hand, Concealed Reverse Grip
   Ready Position 7:   Two Hand, Riot-Ready Thighs
   Ready Position 8:   Two Hand, Ready Hamstrings
   Ready Position 9:   Two Hand, Ready Port Arms
   Ready Position 10: Two Hand, Riot-Ready Thrust
   Ready Position 11: One Hand, Knee-High
   Ready Position 12: Two Hand, Knee-High
   Ready Position 13: One Hand, Grounded
   Ready Position 14: Two Hand, Grounded

Chapter 8: The Combat Clock

The clock points. I am familiar with military terminology and concept from my Army experience. If you were on a foot patrol and the point man suddenly shouted, "enemy at 2 o'clock!" Everyone would instantly look in that direction. The same for pilots - who also have both a vertical clock and a horizontal clock. "12 o'clock high!" Simple. Quick. Effective. Unforgettable.

Yet, scores of differing police and martial arts training systems are not clock-based. Their elaborate weapons angles of attack are disjointed and forgettable, a major problem in this frustrating rat race of systemologies, and the various lines of attack protocols each used for hand, stick, knife and gun tactics. Each try to outsmart or out-do the other, rather than focus on maximizing education. The worst, in my opinion, are the two extremes - the over-simplified and the over-complicated. I began to ask myself, how are all these directions of combat the same? It became clear that attacks universally come in from the center, high or low, or right or left sides, whether standing or on the ground.

No matter the weapon, the angles/directions are the same. I returned with trust to the simple military "combat" clock. The clock face is an imprinted image in our minds since early childhood. The simple angle of attack pattern is right on your wrist, work or play. I discovered, or better stated, I re-discovered the simple, military clock method as a training foundation. Stand it up or lay it down, you have an unforgettable pattern to teach, memorize and work from.

> **Basic Hand, Stick, Knife, Gun Combat Clock Training:**
> 12 o'clock from axis to above
> 3 o'clock from axis to the right
> 6 o'clock from axis to below
> 9 o'clock from axis to the left
> Axis point is the center.

> **Advanced Hand, Stick, Knife, Gun Combat Clock Training:**
> From the axis center point out to all 12 numbers of the clock. This offers more precision training if needed.

Basic Combat Clock Training: 12, 3, 6, 9, Axis

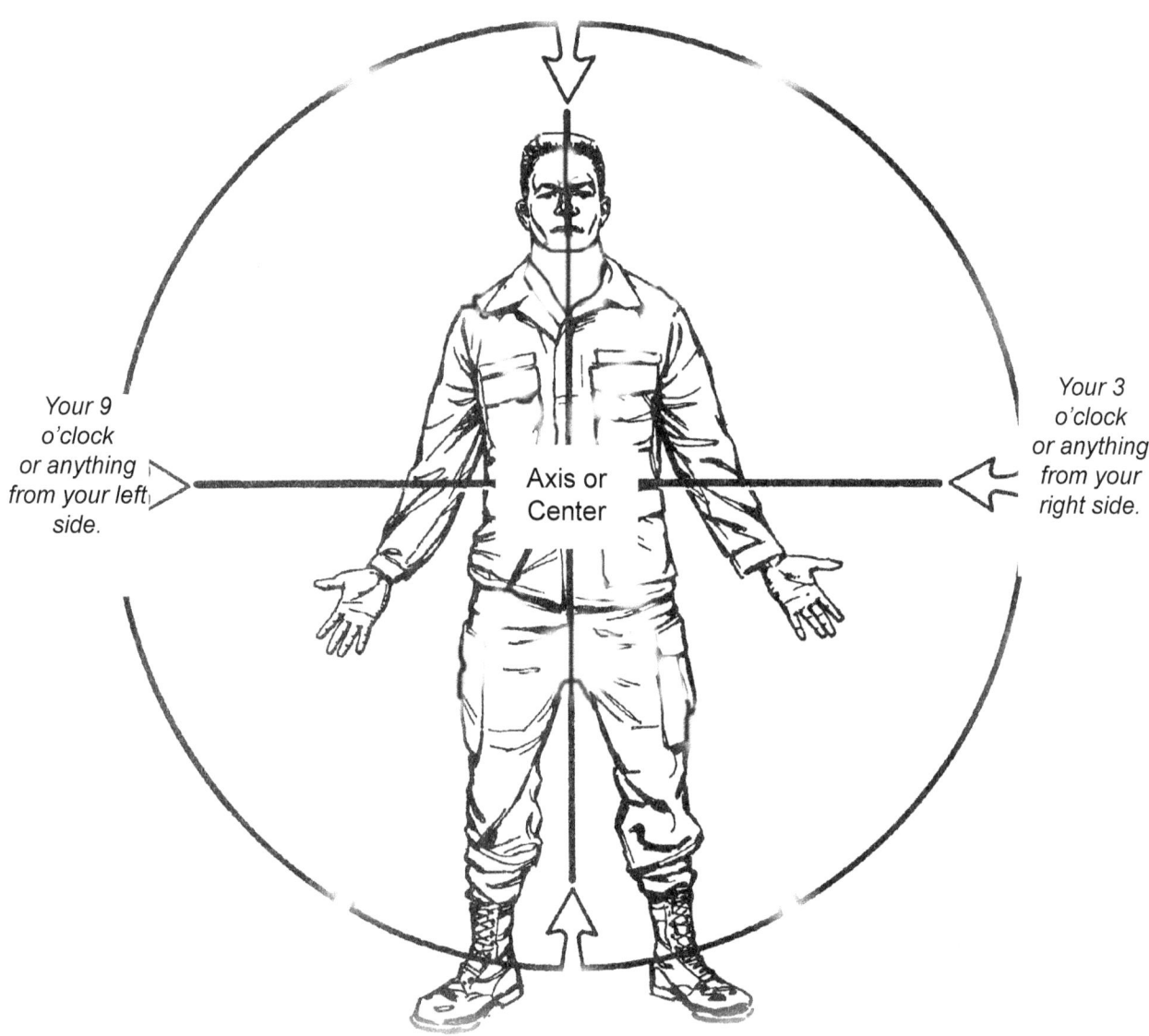

In this Basic Training Format using the four corners and the center of the clock, an instructor can have trainees operating in class and interacting with others quickly and efficiently. As detailed in a subsequent section and summarized here, a trainee will slash or stab his impact weapon from:

        12 o'clock or anything from high.
        3 o'clock or anything from the right side.
        6 o'clock or anything from below.
        9 o'clock or anything from the left side.

Advanced Combat Clock Training: 1 o' clock through 12 o' clock and the Axis.

This Advanced Training Format uses more specific targeting. By targeting the complete 12 numbers of the clock and the center of the clock, an instructor can have trainees operating in class and interacting with others. As detailed in a subsequent section and summarized here, a trainee will slash or stab his impact weapon from:

12 o'clock strike	6 o'clock strike
1 o'clock strike	7 o'clock strike
2 o'clock strike	8 o'clock strike
3 o'clock strike	9 o'clock strike
4 o'clock strike	10 o'clock strike
5 o'clock strike	11 o'clock strike

Note 1: I have taught thousands and thousands of people from utter novices to experts, cadets and rookie cops to martial arts black belts, from all over the world. I can get them to interact with each other in mere moments by using this simple basic clock point format.

Note 2: Remember, there is a difference between target spotting and angles of delivery. When practicing these clock angles alone and "in the air," so to speak, you are only learning "weapon manipulation" skills. Whatever the weapon, be it a hook punch or a stick strike - "Solo Command and Mastery" skills - as I call them, DO NOT assign strikes to body part targets in your official nomenclature. Do not always call a 12 o'clock strike a "head strike," or a 3 o'clock hit a "heart strike." In combat we do not know what position the enemy's body will be in. Plus, your first strike, stab or shot will probably change the template! You are only practicing a directional delivery skills when working alone. If you always imagine hitting specific body parts in your solo workouts, then change up the targets in your mind. Keep your mind open to options, to hunting moving targets that change. A downward 12 o'clock strike might be the face, or to the top of an arm thrusting a punch or stab at you.

Note 3: Sure, we see digital clocks here and there, but we will recall the clock face for another century, maybe longer.

Note 4: I recall in the 1990s, a police academy instructor, kicked back in his office chair, feet on his desk, complaining to me, *"You can't teach these people anything. If you teach them three angles of attack? In six months, they will forget two of them."* Instead of complaining, he should have been judiciously working ways to develop and mold doctrine into unforgettable, high retention methods. Lazy, uninspired bastard. He was wearing a wristwatch, by the way. The Combat Clock. Simple. Timeless. Unforgettable. Versatile.

The Combat Clock as an Attack or Defense Model

The window of combat, as it pertains to the standing, ready, fighting positions, is where you maintain your limbs as a base to move, strike, block and return to. The window of combat is the Combat Clock encircling you and loosely bordered by your shoulders down to your mid-thighs. This "window concept" is especially important in stick vs. stick dueling, a subject covered in-depth in another chapter of this training manual.

Imagine positioning your empty hand at various clock numbers, or your knife at various clock numbers. This is a method of training that has deep potential and runs from the very simple to the very complex, should the training require advanced training such as ambush and assassination criteria.

Using the Combat Clock as an attack or defense module, the instructor can organize groups into specific knife positions and hand positions, such as seen in the photo to the left. These same applications may be made for ground fighting.

The Combat Clock as an Attack or Defense Module. Stick hand at his 2 o'clock line, free, support hand at his 10 o'clock area.

Allow practitioners to improvise a variety of stances and flexibility. The fighting position you choose will be strategically selected to counter the enemy you face before you, and one that will fit into the situation. There is no such thing as one perfect stick fighting stance.

The Combat Clock used as a Attack Defense Model, as applied to ground fighting positions and the position of your enemy.

The Combat Clock is used to:

- learn stick manipulation and solo command and mastery skills.
- maneuvering - organize attack and defense footwork if laid horizontal.
- target spotting - direct fire and locate enemies with a vertical and horizontal clock.
- delivery system - use to deliver angles of attack.
- organize attack striking, hooking/slashing strikes if set vertically.
- organize attack shooting/stabbing/thrusting points if set vertically.
- organize defensive moves if set vertically.
- coordinate mission timing.

Why invent, re-invent and attempt to create and then memorize new, disjointed forgettable, angle systems when the clock and clock face numbers already cover them all, and have been imbedded into our psyche since childbirth?

The Combat Clock for Stick Attack, Stick Defense and Combat Footwork

Chapter 9: Introduction to Footwork

Can you walk?
Can you run?
Can you play basketball or football?

There are basic, universal principles of movement that any combatant must know and apply to successfully defeat an opponent. The principles mentioned are only a few of the basic guidelines that are essential knowledge for stick combat. There are many others which, through years of study, become intuitive to a highly skilled fighter.

a. *Physical Balance.* Balance refers to the ability to maintain equilibrium and to remain in a stable, upright position. A fighter must maintain his balance both to defend himself and to launch an effective attack. Without balance, the fighter has no stability with which to defend himself, nor does he have a base of power for an attack. The fighter must understand two aspects of balance in a struggle:

   (1) How to move his body to keep or regain his own balance. A fighter develops balance through experience, but usually he keeps his feet about shoulder-width apart and his knees flexed. He lowers his center of gravity to increase stability.

   (2) How to exploit weaknesses in his opponent's balance. Experience also gives the hand-to-hand fighter a sense of how to move his body in a fight to maintain his balance while exposing the enemy's weak points.

b. *Mental Balance.* The successful fighter must also maintain a mental balance. He must not allow fear or anger to overcome his ability to concentrate or to react instinctively in close quarters combat.

c. *Position.* Position refers to the location of the fighter (defender) in relation to his opponent. A vital principle when being attacked is for the defender to move his body to a safe position-that is where the attack can not continue unless the enemy moves his whole body. To position for a counter attack, a fighter should move his whole body off the opponent's line of attack. Then, the opponent has to change his position to continue the attack. It is usually safe to move off the line of attack at a 45-degree angle, either toward the opponent or away from him, whichever is appropriate. This position affords the fighter safety and allows him to exploit weaknesses in the enemy's counter attack position. Movement to an advantageous position requires accurate timing and distance perception.

d. Timing. A fighter must be able to perceive the best time to move to an advantageous position in an attack. If he moves too soon, the enemy will anticipate his movement and adjust the attack. If the fighter moves too late, the enemy will strike him. Similarly, the fighter must launch his attack or counter attack at the critical instant when the opponent is the most vulnerable.

e. Distance. Distance is the relative distance between the positions of opponents. A fighter positions himself where distance is to his advantage. The stick fighter must adjust his distance by changing position and developing attacks or counter attacks. He does this according to the range at which he and his opponent are engaged.

f. Momentum. Momentum is the tendency of a body in motion to continue in the direction of motion unless acted on by another force. Body mass in motion develops momentum. The greater the body mass or speed of movement, the greater the momentum. Therefore, a fighter must understand the effects of this principle and apply it to his advantage.

   (1) The fighter can use his opponent's momentum to his advantage-that is, he can place the opponent in a vulnerable position by using his momentum against him.

      (a) The opponent's balance can be taken away by using his own momentum.

      (b) The opponent can be forced to extend farther than he expected, causing him to stop and change his direction of motion to continue his attack.

      (c) An opponent's momentum can be used to add power to a fighter's own attack or counter attack by combining body masses in motion.

   (2) The fighter must be aware that the enemy can also take advantage of the principle of momentum. Therefore, the fighter must avoid placing himself in an awkward or vulnerable position, and he must not allow himself to extend too far.

g. Leverage. A fighter uses leverage in combat by using the natural movement of his body to place his opponent in a position of unnatural movement. The fighter uses his body or parts of his body to create a natural mechanical advantage over parts of the enemy's body.

h. Footwork. Combat footwork is a mixture between sport footwork, as in kick boxing, basketball, football and obstacle course work. The two realms together create the mobile hand, stick, knife and gun combatant.

1: Stationary right foot. Left foot steps forward and back: This is an excellent advance and retreat step that keeps you in range. From a fighting stance, leave your right foot stationary and step forward with your left foot, then step back.

    a) Combat Clock sample - Right foot stays on clock axis. Left foot moves from 10 to 7 o'clock positions.

2: Stationary left foot. Right foot steps forward and back: This is an excellent advance and retreat footwork that keeps you in close range. Do this from a stationary left foot, with the right stepping forward and back. This is advance and retreat footwork.

    a) Combat Clock sample - Left foot stays on clock axis. Right foot moves from 2 to 5 o'clock positions.

3: Shuffle footwork (the pendulum). If you are shuffling forward, let your rear foot come forward near your front foot, displace it, and let your front foot shift forward. Your feet do not have to hit together, or your rear foot does not have to knock the front one forward. But this is sometimes a wise practice for the beginner to first learn the concept. The reverse is used for going backward. This is an exceptionally good move increasing the gap between you and your opponent in a retreat and to advance upon the opponent for delivering many of the low-line kicks conducive to impact weapon fighting. Think of cutting across all the numbers of the clock.

    a) Combat Clock sample - Left foot at 10. Right foot on axis. Left foot comes to the axis. Right foot moves to 4 o'clock.

4: Lunge footwork. Like a fencer, slightly lift your lead foot and propel forward off your rear foot. Do the reverse for going backward. Lift your rear foot and spring back.

5: Lateral footwork. From the fighting stance, if you choose to go a step to the right, then let your right foot step right. Let your left foot follow and move back into stance. If you choose to go to the left, and then let your left foot step first to the left, then let your right foot follow. Then return to the stance. Try not to cross your feet, for this is a point of imbalance.

6: Rocker shuffle. From a fighting stance, slightly bounce your weight on your lead and rear feet shuffling back and forth, or side-to-side. Then change leads in motion and do so from there. Don't over exaggerate the bounce. Stay close to the ground.

7: Sprint forward, back and to the sides with explosive power. For refined work, sprint out of the Combat Clock using all the clock numbers.

8: Back peddling: It is important to be able to back quickly away from a situation. In the gym, learn how not to trip over your own two feet. In the real world you must be careful not to trip over objects.

9: The obstacle course. All of the above mean little without the ability to negotiate a challenging set of real world obstacles and terrain. That's why the elite police and military require their exponents to regularly run and traverse obstacle courses. Many athletic endeavors require similar training. Experiment with this by holding a stick through the course.

10: Now run! Warriors run. They jog; they dash; they hop; they leap; they zig zag; they move through space. Never stop running as long as your legs still work. Run in all kinds of weather. A warrior toughens his or her soul by experiencing discomfort that comes from running. The residual benefits are also vital. Run while holding a training stick, or any tool you will carry!

Remember to use common sports as hyper-links to fighting skills. At times, fighting footwork resembles rugby, basketball, football and soccer. Remind the soldier that he or she already has training in fighting footwork through the sports activities they pursue. Make the connection.

11: *Do not practice barefoot.* Training barefoot is like practicing to ice skate without skates. It just does not work in real combat. Wear what you will really wear when you think you may be in combat or attacked. Barefoot training also leads to more injuries to the ankles, toes and feet. Toughening the feet is an abstract, ancient concept.

12: Lastly, strike and kick while running. Combat is walking and running in explosive seconds while fighting. The soldier should practice all fighting techniques from a run or jog, not always from a static, stance position.

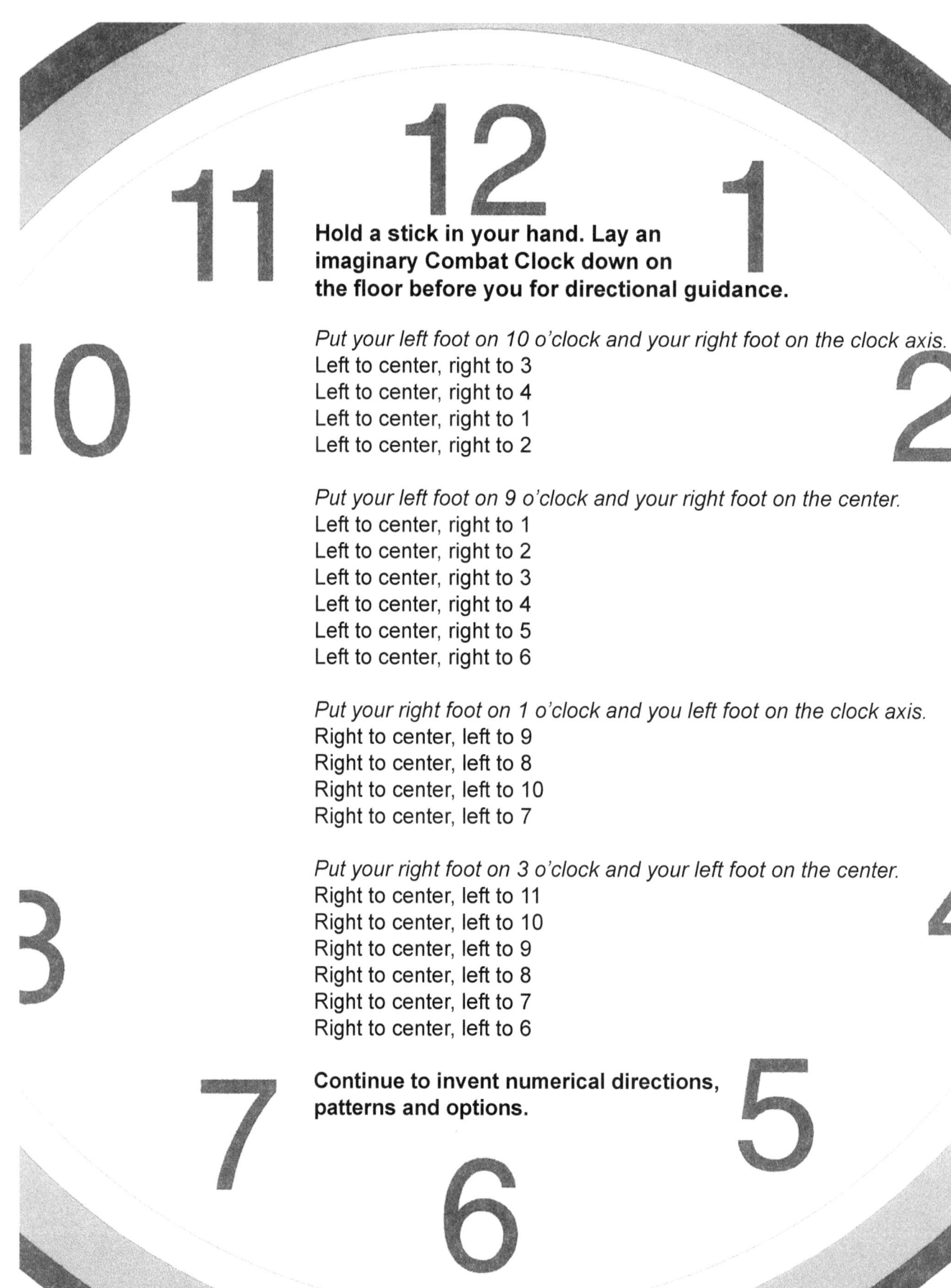

Hold a stick in your hand. Lay an imaginary Combat Clock down on the floor before you for directional guidance.

Put your left foot on 10 o'clock and your right foot on the clock axis.
Left to center, right to 3
Left to center, right to 4
Left to center, right to 1
Left to center, right to 2

Put your left foot on 9 o'clock and your right foot on the center.
Left to center, right to 1
Left to center, right to 2
Left to center, right to 3
Left to center, right to 4
Left to center, right to 5
Left to center, right to 6

Put your right foot on 1 o'clock and you left foot on the clock axis.
Right to center, left to 9
Right to center, left to 8
Right to center, left to 10
Right to center, left to 7

Put your right foot on 3 o'clock and your left foot on the center.
Right to center, left to 11
Right to center, left to 10
Right to center, left to 9
Right to center, left to 8
Right to center, left to 7
Right to center, left to 6

Continue to invent numerical directions, patterns and options.

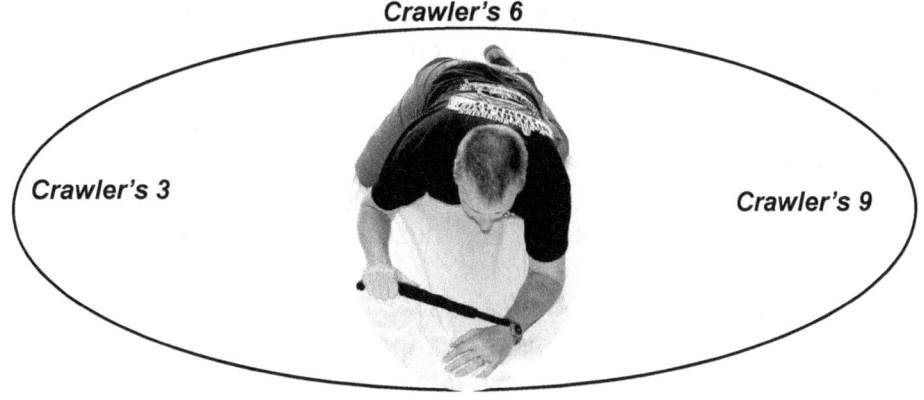

Laid on the ground, the Combat Clock can be utilized for exercising all directions of crawling.

    Crawl Exercise 1: The soldier lays flat on chest and, on command of a number, posts/jumps up and bolts off the clock in the direction of that number.

    Crawl Exercise 2: The soldier lays flat in the center. On command, he crawls off the clock in the direction of the clock number.

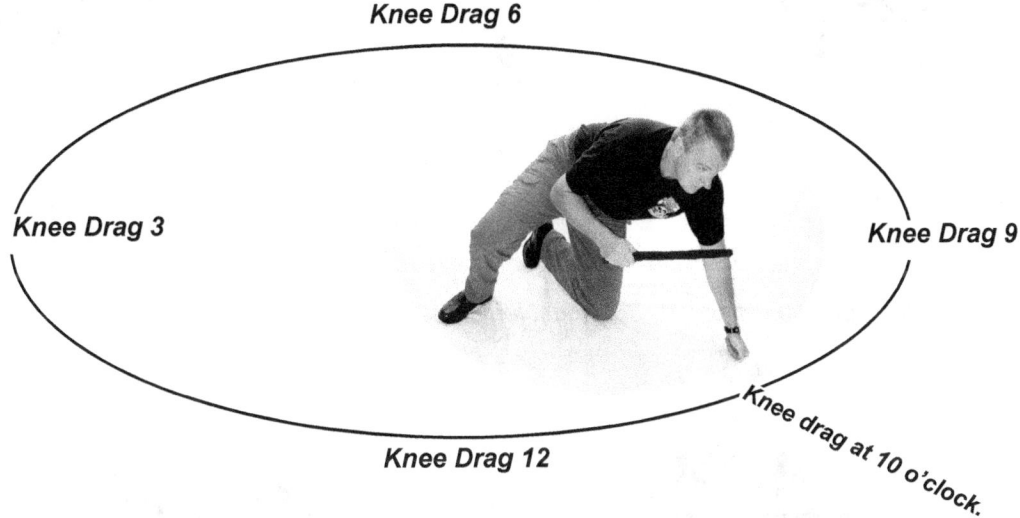

Laid on the ground, the Combat Clock can be utilized for exercising all directions of knee drags with a stick. A knee drag is an old military exercise that soldiers use when closing in on an enemy. The fist, or elbow/forearm post moves, then the posted knee follows when traveling forward on the clock.

    Knee Drag Exercise 1: The soldier "drags" to a called out number on the clock.

Knee High Maneuver Drills "While Holding" an Impact Weapon.
In close quarter combat, it is imperative to be confident and comfortable maneuvering knee high. If given an opportunity to do so, remaining knee high is usually as low as you should get if an opponent is on his back or chest. Knee high is the superior ground fighting position. It is where support hand, forearm, elbow and knee strikes are easily delivered. Knee high is also the preferred arrest position. Real world combatives are not high school, college and sport judo matches. Of course, you must study these sports also to learn their tricks and calculate how to defeat the tactics.

In order to enhance these vital knee maneuvering skills, I have organized this knee high drill which I teach with empty hands, impact weapons, firearms and knives. For the "Knee High" drill, imagine a clock laid down on the floor, and you are at its center, down on both knees. You will start by stepping right foot, then left foot through the progression.

The center, neutral position

Right leg steps out to front

Return to the center neutral.

Left leg steps out.

Return to the center neutral.

Right foot to 1:30.

Return to middle.

Then extend left foot to 10:30.

Return to the middle.

Right foot out to 3 o'clock.

Return to the middle.

Left foot out to 9 o'clock.

Back to the neutral.

Right foot to 4:30.

Return to the middle.

Left foot out to 7:30.

Return to the middle.

Right foot back to 6 o'clock.

Return to the middle.

Left foot back to 6 o'clock.

And return to neutral.

CHAPTER 10: IMPACT WEAPON STRIKES

Part 1: Closed Baton/Short Stick Tactics

Civilians, officers, agents and members of the military often find themselves in violent circumstances where they have drawn their closed baton from their carry site under great stress and do not have a spare second to actually open it. They must use the closed baton in their hand like a very short stick. Further, upon being properly trained in these short stick tactics, personnel might also opt to keep their baton collapsed for the many useful methods such a shortened tool offers.

Working with the closed tool often relieves the stigma of brandishing the fully opened, longer-length, impact weapon. The closed baton still acts as a close quarter impact weapon, striking with the top-side and bottom-side of the baton that extends outside the top and bottom of your closed fist. For training doctrine, it is wise to label the top and bottom sides for proper practice. The top side of the extended baton should be called the "saber side." The bottom side should be called the "reverse side." These labels work in conjunction with all other tactics and makes for easier learning and memory.

These exposed ends may also act as short, temporary tools for hooking and grappling. This will depend on the length of the closed baton handle. Some handles are manufactured very short, and others are quite long.

Plus, this closed baton may act as hand support for power punching. Your hand must wrap tightly around the girth of the baton handle. Strike training objects in this manner for a test. If this feels uncomfortable, do not use it as punching support, as hand size and baton handle girth will vary. If it fits well, you now have a new use for this popular weapon with the common jab, the cross, the hook, the uppercut and the overhand strike. Holding a properly sized, solid object in your hand reinforces the structure of your fist.

The Short Stick

Police agencies, self defense citizens, criminals and some martial arts around the world use a small stick like the Korean Dan Bong or the truncheon. This small stick can be the size of a big, yet closed expandable baton. It can be loosely defined as approximately 6, 7 or 8 inches to 12 or 14 inches long. Many of the closed baton tactics can work with the short stick. Many of the longer stick tactics can also work with a short stick. Keep in mind grappling with a short stick has its "shortcomings."

Many grappling moves are shown in martial arts, and/or transferred to short flashlights. I don't trust most of these moves.

Hansons' military expert Adrian Stevenson poses with a selection of British Police Truncheons from the collection of Dr. Michael Carter.

I ask you to experiment with short stick moves from the longer stick moves that I show here in this book. This way I do not have to replicate *every* long stick photo with a short stick photo version that has the same motion.

It is wise to label the exposed top as the "saber side" and the exposed bottom as the "icepick'"or "reverse side".

The Closed Baton and Small Stick Ready Positions
You might transition to and through any of these positions as all fighting is situational. It is wise to label the portion of the weapon that protudes from the top of the hand as "saber side." Then label the bottom portion as "reverse grip," or labeled by some as "icepick side"). This defines and maximizes training methods and limits confusion.

Closed baton or short stick, you will be:
    Thrust striking with the saber side.
    Hook striking with the saber side.
    Thrust striking with the icepick or reverse side.
    Hook striking with the icepick or reverse side.
    Thrusting punches.
    Hooking punches.
    Limited blocking.
    Limited grappling.

If there is enough top and/or bottom exposure, the weapon may be used as a hook or grappling tool.

If your hand fits well with the girth of the weapon, you may use the closed baton in support when punching.

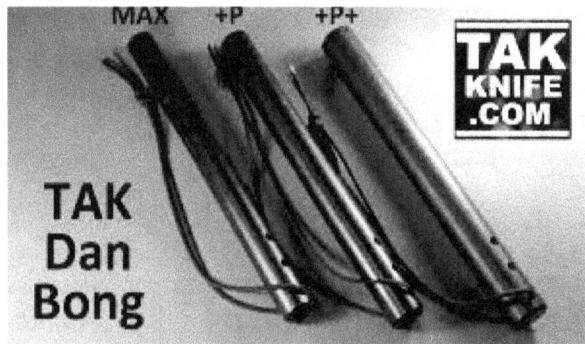

1: The weapon up front.

2: The weapon neutral.

3: The weapon to the rear.

4: Knee high (fighting high).

Fghting equal height.

Fighting below, or the topside of a ground fight.

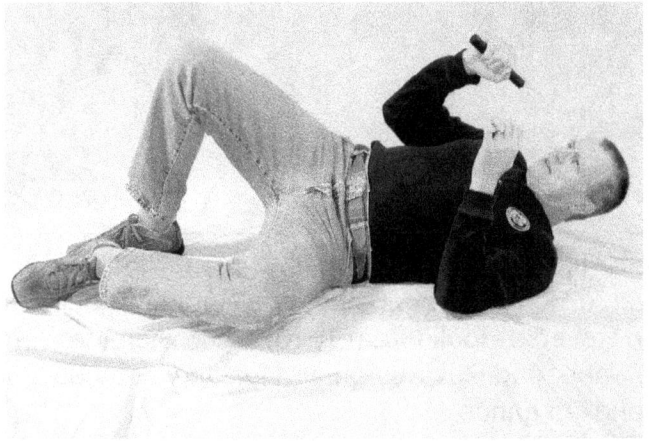
5: Fighting from your back and sides.

The Closed Baton and Short Stick Hooking Strikes

1: Hook from 12 o'clock or anything from above.

2: Hook from 3 o'clock or anything from the right.

3: Hook from 6 o'clock or anything from below.

4: Hook from 9 o'clock or anything from the left.

5a: The X Hook from 2 o'clock.

5b: The follow-up X Hook from 10 o'clock.

Basic Training, Saber Hooks 12, 3, 6, 9 on the Combat Clock.
Basic Training, Reverse Hooks 12, 3, 6, 9 on the Combat Clock.
Basic "X" Saber and Reverse Hooks from 1, 10, 4 and 8 on the Combat Clock.
Advanced Training, Saber and Reverse on all 12 of the Combat Clock numbers.
- work right and left hands.
- work standing, walking, kneeling and on your back and sides.
- hit training objects solo as on bags and gear held by a trainer.

The Closed Baton and Short Stick Thrusting Strikes

1: A 12 o'clock or high to anywhere thrust.

2: A 3 o'clock or right side thrust.

3: A 6 o'clock low to anywhere thrust.

4: A 9 o'clock or left side thrust.

Reverse strikes (like backhanded) on 12, 3, 6 and 9.

Basic Training, Saber Thrusts 12, 3, 6, 9 on the Combat Clock.

Basic Training, Reverse Thrusts 12, 3, 6, 9 on the Combat Clock.

Advanced Training, Saber and Reverse on all 12 of the Combat Clock numbers.

- work right and left hands.
- work standing, walking, kneeling and on your back and sides.
- hit training objects solo as on bags and gear held by a trainer.

The 5 Punch Closed Baton Set This exercise drill works punching "while holding" the closed baton, or short stick. A practitioner must cycle through right and left leads, right and left hand grips in four sets. The 5 common strikes are jab, cross, hook, uppercut, descending overhand. Switching hands and leads covers all possibilities.

1: Right lead, right, "jab" punch while holding baton.

2: Right lead, "cross" empty hand punch.

3: Right lead, right "hook" punch while holding baton.

4: Right lead, "uppercut" empty hand punch.

5: Right overhand "descending" punch.

Remember that some closed batons may accidentally open while you thrust strike, hook strike or punch with them. This can be advantageous at times. Do not be surprised and "stall out" if this happens.

Exercise the 5:
The set is-
Jab, cross, hook, uppercut, descending overhand

Right hand holds the weapon.
Left hand holds the weapon.
Switch lead legs.
Add in any kicks.

- work standing, walking, kneeling and on your back and sides.
- hit training objects solo as on bags and gear held by a trainer.

The Closed Baton and Small Stick Workout

Thrusting Saber
    Combat Clock Basic 4 and Advanced 12.
    Right and left hand.
    Set your own number of reps.

Thrusting Reverse
    Combat Clock Basic 4 and Advanced 12
    Right and left hand.
    Set your own number of reps.

Hooking Saber
    Combat Clock Basic 4 and Advanced 12.
    X pattern.
    Right and left hand.
    Set your own number of reps.

Hooking Reverse
    Combat Clock Basic 4 and Advanced 12
    X pattern
    Right and left hand
    Set your own number of reps.

Punching
    Jab, cross, hook, uppercut overhand.
    Right hand holds weapon.
    Left hand holds weapon.
    Add in any kicks.
    Set your own number of reps.

Perform all
    Standing.
    Walking.
    With kicks.
    Knee high.
      - to high.
      - to equal height.
      - to low, or topside of a ground fight.

    Grounded on back and side.
      - add ground kicks.

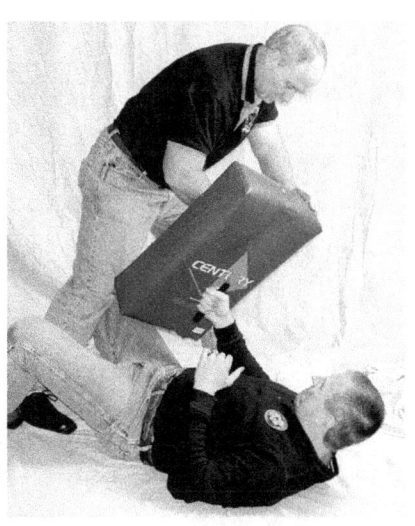

Closed Baton Combatives Practice Through the Stop 6
Exercise the stress quick draw and closed baton strikes through each situation in the Stop 6. The practitioner should freestyle responses until the conflict seems at an appropriate end for the situation. This module is only meant to experiment and exercise the closed impact weapon strikes. Subsequent actions will be studied in other levels. Obviously, this module can be used in regular classes to develop all kinds of follow-ups and combat scenario skills.

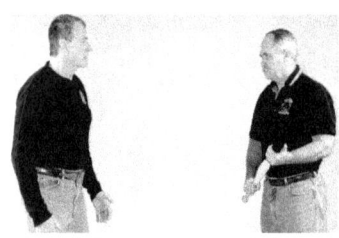

Stop 1: The Standoff, Showdown Stop
Stop 1: Closed weapon striking, supported by empty-hand strikes and kicks.

Stop 2: Hands-on Stop
Free your fingers and hands with releasing techniques.
- circular and joint lock releases.
- yank-outs, slap releases.
- other tricks.

Stop 2: Release from grip, draw and barrage the enemy.

Stop 3: Forearm to Forearm Stop
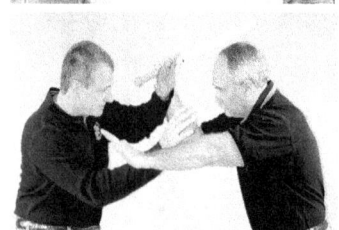

Stop 3a: Draw, unleash a barrage of strikes and kicks.
Stop 3b: Learn the optional *Block, Pass and Pin Series.*
Stop 3c: Learn optional *Outside Invasion Series.*
Stop 3d: Learn the optional *Critical Contact Series.*

Stop 4: Biceps-Neck-Biceps Stop

Stop 4: Draw, unleash a barrage of strikes and kicks.

Stop 5: The Bear Hug, Clinch Stop

Stop 5: Draw, unleash a barrage of strikes and kicks.

Stop 6: The Grounded Stop
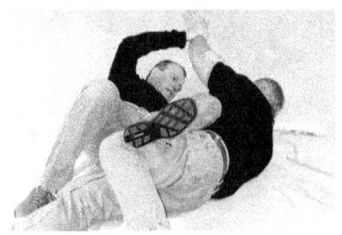

Stop 6: Get to a position, via striking, kicks and grappling. Draw, unleash a barrage of strikes and kicks.

Page 68

Part 2: The Longer Stick/Baton and Expanded Impact Weapon Strikes

Here are some of the foundational aspects of hitting with a regular-sized stick, suitable for one hand operation, yet not in the small stick category.

Impact Weapon Striking Overview

The Four Grips:
  Grip 1: one hand on one end.
  Grip 2: two hands spread apart.
  Grip 3: one hand in the center.
  Grip 4: the batting or sword grip.

The Impact Weapon Strikes With:
  Strike 1: the shaft.
  Strike 2: the tip.
  Strike 3: the handle.
    a) side of the handle.
    b) pommel.

The Impact Weapon Strikes:
  Delivery 1: with stabbing thrust.
  Delivery 2: with hooking angles.
  Delivery 3: with committed lunge, follow-throughs.
  Delivery 4: with hit and retracts.
  Delivery 5: from the Combat Clock numbers/angles.
  Delivery 6: with a synergetic, body support position.
  Delivery 7: with a single-hand grip.
  Delivery 8: with a double-handed grip.

The Impact Weapon is Helped With:
  Help 1: Support arm strikes.
  Help 2: Support leg kicks.
  Help 3: Support body rams.

The shaft/barrel strike.

The tip stab.

The pommel strike.

Body Position:
- Keep the slash economical.
- Keep the action in the window of combat.
- Keep the support hand up and ready.
- Keep a good, athletic balance.

Single Hand Shaft Strikes:
- The slashing saber grip.
  - single slash.
  - double or more slashes.
  - X or figure 8 slashes.
- The hit and retract shaft strikes.
- The reverse grip slashes.
- The reverse grip hit and retract.
- The fanning strikes.
- The circular strikes.
  - big circles.
  - small circles.

Single Hand Tip Stabs:
- Hooking stabs.
- Thrusting stabs.

Single Hand Pommel Strikes:
- Hooking hits.
- Thrusting hits.

Two Hand Shaft Strikes:
- Hooking strikes.
- Thrusting strikes.

Two Hand Bayonet or Front Tip Strikes:
- Hooking hits.
- Thrusting hits.

Two Hand Stick or Rifle Butt Rear Tip Strikes:
- Hooking hits.
- Thrusting hits.

Single Hand Stick Strikes Overview
The single-hand, saber grip slash is probably the most commonly utilized application. Exercised in the air, the slash has a complete follow-through. When actually hitting something, the weapon may stop on the enemy, delivering a fluid shock in the area of the strike. Once temporarily stopped on impact, for a slash to continue, the stick needs to be drawn back just a bit, drawn from the target site to clear the target for the stick to continue its so-called "slash." Also the target may shatter or collapse away on impact, and the slash may continue.

The stop on impact is so fast it doesn't appear to be a stop at all. Why continue with the slash? You may wish or need to strike again, back from the opposite side.

The wrist is a key pivot point for power.

The Power of the Common Swing
The most common stick attack is the single-hand, swinging motion using the shaft as the striking surface. How much power do you need? How much power can you generate?

Earlier in this book we covered the hand grip and choking up on the stick as it relates to speed and power. As you recall, there will be some lingering debate amongst the dogmatic about where to grip the stick, but most practitioners choke up on the stick. The common mechanics of a stick swing also has some debated theories. In general terms, there is the martial arts stick swing and the police baton swing.

The elbow is a key pivot point for power.

The Martial Arts Swing
The martial arts swing will dictate that the practitioner use more of the wrist, more of the elbow and minimize the use of the shoulder. In fact, too much rearing back, or "chambering" of the stick will allow the opponent time to read and react.

The second part of this is the classic overswing. If the practitioner overswings, then this also allows the opponent an opportunity to charge in and counter your attack. In Filipino Martial Arts, for example, well-known for their stick expertise, this over-chambering and over swinging as part of an invasion and footwork drill for both stick-versus-stick fighting and unarmed-versus-the-stick. It is a truth that common people will over chamber and over swing.

The shoulder is a key pivot point for power, but many believe not to rear your weapon back too far.

So, the martial arts stick swing will be very efficient and delivered with more of an arc than a half-circle. Go for little high end chambering in the beginning and no over swing at the end.

Experts warn against the overswing from the shoulder. The enemy can rush you at this point.

The Police Baton Swing

The standard, universal, stick swing of law enforcement, security and the military never considers the high chamber or the overswing as a weakness. In fact, enforcement training mandates that officers stand with the stick virtually posed way back on their shoulder as a ready position. The police experts explain power development from the wrist through the shoulder and include body movement. They start with explaining a wrist flick that generates a little power. Then they add the elbow movement adding more power, then ask for dynamic shoulder involvement in the swing for maximum power. Further, they add the twisting torso for added force, also mentioning a well-balanced stance. Finally, they describe the whole body descending in with the swing to add the last bit of power.

Military and police experts report that under combat stress, while targeting specific points on the body, 40 percent of the time your strike will be high, 40 percent your strike will be low, 10 percent of the time you will miss, and 10 percent of the time you hit as planned. This is called by Dr. Kevin Parsons of American Systems and Procedures as the *40/10 Theory*. These stats support the four quarters of the clock striking approach. It is a quick study method where you simply strike down, from the right, the left and upward in general and powerful attacks. More time to practice should produce better targeting results. Classify your time, your goals, yourself and your students as the beginner, intermediate or advanced.

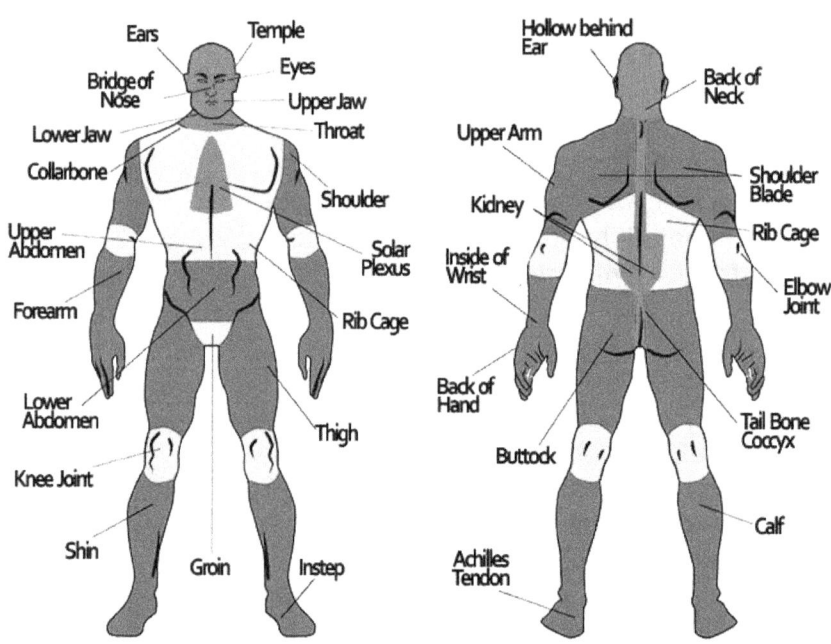

STRIKING AREAS

GREEN	YELLOW	RED
REASONING	REASONING	REASONING
Minimal level of resultant trauma. Injury tends to be temporary rather than long-lasting, however exceptions can occur. Except for the HEAD, NECK, SPINE, the whole body is a Green Target Area for the application of baton blocking and restraint skills.	Moderate to serious level of resultant trauma. Injury tends to be more long-lasting, but may also be temporary.	Highest level of resultant trauma. Injury tends to range from serious to long-lasting rather than temporary and may include unconsciousness, serious bodily injury, shock or death.

Single Hand, Single Saber Slash (Hooking) Shaft Strikes

12 o'clock or high to anything lower slash.

3 o'clock or right side slash inward.

6 o'clock or low to anything higher slash.

9 o'clock or left side to a backhand slash.

Basic Training, Saber Slash (Hooks) 12, 3, 6, 9 on the Combat Clock.
Basic "X" Saber Slash (Hooks) from 2, 10, 4 and 8 on the Combat Clock.
Advanced Training, Saber Slash (Hooks) on all 12 of the Combat Clock numbers.
- work right and left hands.
- work standing, walking, kneeling and on your back and sides.
- hit training objects solo as on bags and gear held by a trainer.

Note: Do the above with a long and short stick.

Single Hand, Double Saber Slash (Hooking) Shaft Strikes

Down and up (12 and 6 o'clock strikes).

Inward and backhand (3 and 9 o'clock strikes).

Up and down (6 and 12 o'clock strikes).

Backhand and inward (9 and 3 o'clock strikes).

Basic Training, Saber Slash (Hooks) 12, 3, 6, 9 on the Combat Clock.
Basic "X" Saber Slash (Hooks) from 2, 10, 4 and 8 on the Combat Clock.
Advanced Training, Saber Slash on all 12 of the Combat Clock numbers.
- work right and left hands.
- work standing, walking, kneeling and on your back and sides.
- hit training objects solo as on bags and gear held by a trainer.

Note: Do the above with a long and short stick.
Note: Triple threat? Make a third return strike.

Saber X Strikes (2 Slashes) Two Slashes but this time NOT on the same lines.
Understanding the X pattern and its multitude of uses in swinging a stick is vital. It comes from all sides, both high and low and very quickly. It is both an "X" and a figure 8 in 2 slashes.

Barnhart starts the X at 1:30.

Barnhart starts the X at 10:30.

Barnhart starts the X at 4:30.

Set 1 practice - 1:30 then 10:30
Set 2 practice - 10:30 then 1:30
Set 3 practice - 4:30 then 7:30
Set 4 practice - 7:30 then 4:30

The versatility of the X continues. Try tighter angles like 2:30 or 8:30. Or start at 2:30 and return at 11. Remember that an X is any return that does not come back on the same line as the X originated on. That would be just a double slash in our definitions.

Saber Slash X Basic Training
 X1: Start an X from high right.
 X2: Start an X from high left.
 X3: Start an X from low right.
 X4: Start an X from low left.

Repeat with the opposite hand.
Standing (moving).
Exercise these knee high.
 - right knee up.
 - left knee up.
 - both knees down.

Exercise these on the ground.
 - on back.
 - on sides.

Single Hand, Circular Saber Grip Shaft Strikes

There are two kinds of big circles involving the shoulder and small circles involving mostly the wrist, the forearm and then the elbow. These are double or more hits. They are difficult to demonstrate in photos. Key to executing them both is a proper wrist rotation and palm positioning. You will turn the palm either up or down in the proper place.

Big Circles on the Combat Clock

Big Circle 1: A downward striking circle starting from above. The actual motion is more like the shape of an oval racetrack than a complete, perfect circle hinging from the shoulder.

Big Circle 2: A right to left striking circle that encompasses the head. Martial artists will bicker about how close to the head the stick must pass, but it really is a moot issue. Swing where you have to swing.

Big Circle 3: A upward striking circle starting from below. The actual motion is more like the shape of an oval racetrack than a complete, perfect circle hinging from the shoulder.

Single Hand, Circular, Double Shaft Strikes

Big Circle 4: A left to right striking circle, which encompasses the head. Martial artists will bicker about how close to the head the stick must pass, but it really is a moot issue. Swing where you have to swing.

> Basic Training, Saber Circle Slash (Hooks) 12, 3, 6, 9 on the Combat Clock.
> Advanced Training, Saber Slash on all 12 of the Combat Clock numbers.
> - work right and left hands.
> - work standing, walking, kneeling.
> - hit training objects solo as on bags and gear held by a trainer.
>
> *Note: Triple threat? Make a third circle.*
> *Note: Do with both a long and short stick.*

Small Circles on the Combat Clock

You can spin small circles at the wrist both inside the arm and outside the arm. These are meant to be double or more hits, so maintain as solid a grip as possible on the handle.

> Basic Training, Saber Circle Slash (Hooks) 12, 3, 6, 9 on the Combat Clock.
> Advanced Training, Saber Slash on all 12 of the Combat Clock numbers.
> - work right and left hands.
> - work standing, walking, kneeling.
> - hit training objects solo as on bags and gear held by a trainer.
>
> *Note: Triple threat? Make a third circle.*
> *Note: Do with both a long and short stick.*

Single Hand, Saber Grip, Hooking Pommel Strikes

12 o'clock or high to low.   *3 o'clock or right side to left.*

6 o'clock or low to high hook.   *9 o'clock or left side to right side.*

Basic Training, Saber on the Combat Clock.
Basic "X" Saber Slash on the Combat Clock.
Advanced 12 Combat Clock numbers.
- work right and left hands.
- work standing, walking, kneeling and on your back and sides.
- hit training objects solo as on bags and gear held by a trainer.

Note: Do the above with a long and short stick.

Single Hand, Saber Grip, Thrusting Pommel Strikes

12 o'clock or high thrust.

3 o'clock or right side thrust.

6 o'clock or low thrust target.

9 o'clock or left side thrust.

Basic Training, Saber Grip on the Combat Clock. Advanced 12 Training Combat Clock numbers.
- work right and left hands.
- work standing, walking, kneeling and on your back and sides.
- hit training objects solo as on bags and gear held by a trainer.

Note: Do the above with a long and short stick.

Single Hand, Saber Grip, Hooking Tip Stab Strikes

12 o'clock or high hook stab.

3 o'clock or right side hook stab.

6 o'clock or low hook stab.

9 o'clock or left side hook stab.

Basic Training, Saber Stab (Hooks) 12, 3, 6, 9 on the Combat Clock.
Advanced Training, Saber Stab (Hooks) on all 12 of the Combat Clock numbers.
- work right and left hands.
- work standing, walking, kneeling and on your back and sides.
- hit training objects solo as on bags and gear held by a trainer.

Note: Do the above with a long and short stick.

Single Hand, Saber Grip, Thrusting Tip Stab Strikes

12 o'clock or high thrust stab.   *3 o'clock or right side thrust stab.*

6 o'clock or low thrust stab.   *9 o'clock or left side thrust stab.*

Basic Training, Saber Tip Thrust Stab 12, 3, 6, 9 on the Combat Clock.
Advanced Training, Saber Tip Thrust Stab on all 12 of the Combat Clock numbers.
- work right and left hands.
- work standing, walking, kneeling and on your back and sides.
- hit training objects solo as on bags and gear held by a trainer.

Note: Do the above with a long and short stick.

Single Hand, Saber Grip, Fanning Tip Stab Strikes

This strike attacks in the very manner of its name. It fans at the wrist and strikes opposite sides as quickly as possible. The second strike, the actual fan, could be the same height, or higher or lower.

Fan 1: Start with a high 12 o'clock strike, and rotate the wrist to a low 6 o'clock strike. "Head-groin."

Fan 2: Start with a 3 o'clock side strike, and rotate the wrist to a 9 o'clock strike. "Right-left."

Fan 3: Start with a high 6 o'clock strike, and rotate the wrist to a high 12 o'clock strike. "Groin-head."

Fan 4: Start with a 9 o'clock side strike, and rotate the wrist to a 3 o'clock strike. "Left-right."

The trainer can hold a vertical or horizontal striking stick.

Basic Training, Saber Tip fanning Strikes 12, 3, 6, 9 on the Combat Clock.
Advanced Training, Saber Tip Fanning Strikes on all 12 of the Combat Clock numbers.
- work right and left hands.
- work standing, walking, kneeling and on your back and sides.
- hit training objects solo as on bags and gear held by a trainer.

Single Hand, Reverse Grip, Slashing Shaft Strikes

In some systems, this grip is popular as it resembles a walking cane The strike is popular with "cane people."

12 o'clock or from above. From the right or 3. From below or 6-ish. From the left or backhanded.

Basic Training, Reverse Grip Hook 12, 3, 6, 9 on the Combat Clock.
Advanced Training, Reverse Grip Hook on all 12 of the Combat Clock numbers.
- work right and left hands.
- work standing, walking, kneeling and on your back and sides.
- hit training objects solo as on bags and gear held by a trainer.

Note: Do the above with a long and short stick.

Single Hand, Reverse Grip, Thrusting Pommel Strikes

12 o'clock or from above. From the right or 3. From below or 6-ish. From the left or backhanded.

Basic Training, Reverse Grip Thrust 12, 3, 6, 9 on the Combat Clock.
Advanced Training, Reverse Grip Thrust on all 12 of the Combat Clock numbers.
- work right and left hands.
- work standing, walking, kneeling and on your back and sides.
- hit training objects solo as on bags and gear held by a trainer.

Note: Do the above with a long and short stick.

Two Handed Grip Strikes: Better Understanding of The Two Handed Grip

With a two handed grip we strike with the shaft and the two pommels or two ends, with thrusts and hooks. Trying to differentiate between the two pommels, the right and left sides can be a challenge for system doctrine. A very successful way is to consider the stick like a long gun. Consider one end is the barrel or bayonet end, the other is the stock. This helps label and organize proper mind-set and training.

Two Hand Grip Impact Weapon

Strike with:
- the shaft.
- the right end.
- the left end.

In:
- thrusting motions.
- hooking motions.
- committed lunges.
- hit and retracts.

Held:
- both palms down.
- both palms up (rare).
- left palm up, right down.
- left palm down, right up.

Using a long gun as a model tool helps define two handed stick work.

Two Handed Grip, Shaft Thrust Strikes

12, 3, 6 and 9 thrusts.

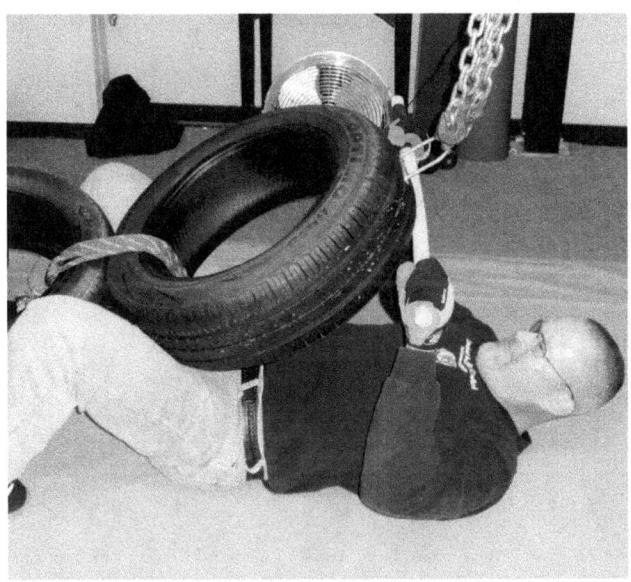

Basic Training, Two Hand Grip Thrust 12, 3, 6, 9 on the Combat Clock.
Advanced Training, Two Hand Grip Thrust on all 12 of the Combat Clock numbers.
- work standing, walking, kneeling and on your back and sides.
- hit training objects solo as on bags and gear held by a trainer.

Two Handed Grip, Shaft Pulling Strikes

Put the stick out, grab with two hands and pull in.

Yank/pull in high.   Yank/pull in from the right.   Yank/pull in low.   Yank, pull in from the left.

Feed the pommel and row back/pull. Feed the right and left sides.

Basic Training, Two Hand Grip Pull 12, 3, 6, 9 on the Combat Clock.
Feed the pommels, feed left, feed right and "row back."
Advanced Training, Two Hand Grip pull on all 12 of the Combat Clock numbers.
- work right and left hands.
- work standing, walking, kneeling and on your back and sides.
- hit training objects solo as on bags and gear held by a trainer.

Two Handed Grip, Shaft Hooking Strikes

12, 3, 6 and 9 hooks, up, down, across left, across right.

Basic Training, Two Hand Grip Hook Strike 12, 3, 6, 9 on the Combat Clock.
Advanced Training, Two Hand Grip Hook on all 12 of the Combat Clock numbers.
- work right and left hands.
- work standing, walking, kneeling and on your back and sides.
- hit training objects solo as on bags and gear held by a trainer.

Two Handed Grip, Tip Slash/Hooking Strikes (Think Bayonet Slash)

Slash down. *Slash from right to left.* *Slash up.* *Slash from left to right.*

Basic Training, Two Hand Grip Tip Slash 12, 3, 6, 9 on the Combat Clock.
X Tip Slashing on 2, 10, 4 and 8 o'clock-ish angles.
Advanced Training, Two Hand Grip, Tip Slash on all 12 of the Combat Clock numbers.
- this is with a rifle grip and work right and left hand leads.
- work standing, walking, kneeling and on your back and sides.
- hit training objects solo as on bags and gear held by a trainer.

Two Handed Grip, Tip Stab Thrust Strikes (Think Bayonet Stab)

Stab mid to high.   *Stab to the right side.*   *Stab mid to low.*   *Stab to the left side.*

Basic Training, Two Hand Grip Tip Stab 12, 3, 6, 9 on the Combat Clock.
Advanced Training, Two Hand Grip, Tip Slash on all 12 of the Combat Clock numbers.
- this is with a rifle grip and work right and left hand leads.
- work standing, walking, kneeling and on your back and sides.
- hit training objects solo as on bags and gear held by a trainer.

Note: Hand grips are spread apart.

Two Handed Grip, Stock-Butt Thrust Strikes (Think Rifle Stock)

Lift into position, 12 o'clock or high thrust.   *Lift into position, 3 o'clock or right side stab.*

Lift into position, 6 o'clock or low thrust.   *Lift into position, 9 o'clock or left side stab.*

Basic Training, Two Hand Stock Thrust 12, 3, 6, 9 on the Combat Clock.
Advanced Training, Two Hand Grip, Stock Thrust on all 12 of the Combat Clock numbers.

- this is with a rifle grip and work right and left hand leads.
- work standing, walking, kneeling and on your back and sides.
- hit training objects solo as on bags and gear held by a trainer.

Two Handed Grip, Slash Hooking Strikes (Think Sword or Batting)

12 from above. *3 from the right.* *6 from below.* *9 from the left.*

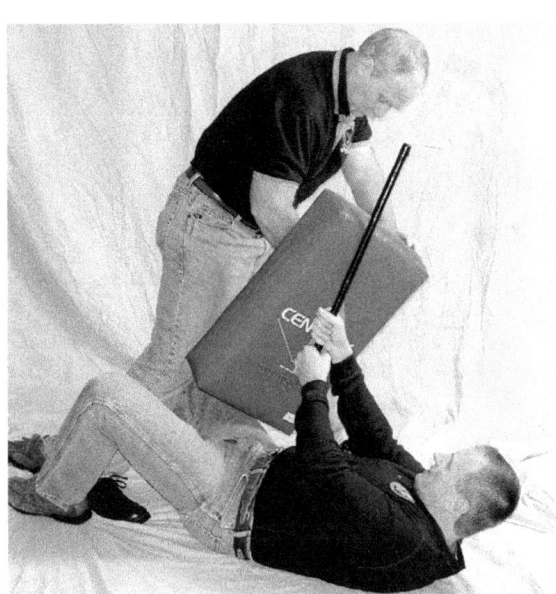

Basic Training, Two Hand "Sword" Slash/Hook 12, 3, 6, 9 on the Combat Clock.
X Pattern from 10, 2, 4 and 8 o'clock.
Advanced Training, Two Hand Grip, Slash on all 12 of the Combat Clock numbers.

- this is with a sword-like grip and work right and left hand leads.
- work standing, walking, kneeling and on your back and sides.
- hit training objects solo as on bags and gear held by a trainer.
  Note: Hand grips are close together on one end.

Two Handed Grip, Stabbing Strikes (Think Sword Stabbing or Batting)

12 or mid to high.   3 or from the right side.   6 or mid to low.   9 from the left side.

WW II, old Fairbairn, how-to diagrams.

Basic Training, Two Hand "Sword" Stab 12, 3, 6, 9 on the Combat Clock.
Advanced Training, Two Hand Grip, Stab on all 12 of the Combat Clock numbers.

- this is with a sword-like grip and work right and left hand leads.
- work standing, walking, kneeling and on your back and sides.
- hit training objects solo as on bags and gear held by a trainer.
  Note: Hand grips are close together on one end.

Two Handed Grip, The Surrounded Series
This is a mere six step, directional, spacial recognition exercise that reminds practitioners that you must be aware of all sides, up and down.

Forward thrust.   *Rear thrust.*   *Left side thrust.*

Right side thrust.   *Upward thrust.*   *Downward thrust.*

Do this right handed (imagine a right handed, rifle trigger finger.)
Do this left handed (imagine a left handed, rifle trigger finger.)
    - look into the direction that you are striking. Try to look before you "leap."

The Statue Drill for Contact and Striking Skill

Throughout the *Force Necessary* courses we use the so-called statue drill. This drill is the foundation for understanding the *point of contact* in combat. New students must be introduced to this full, basic training spectrum. The drill ranges from a solid statue, to the trainer taking a few primary attack steps, all in preparation for the student to face chaotic reality.

Many systems incorporate wooden posts or dummies for this type of training, which has advantages, but all too often training sessions consisting of larger groups will not have such abundant equipment. A live training partner *dummy* provides a more suitable trainer.

Statue Drill Studies and Observations 1) Contact

Contact may come from your strike that is blocked, or from your block defending against a strike. The contact point is a reference point for training. Contact points in an impact weapon encounter may clash:

Clash 1) Your stick to his weapon, such as knife or stick.
Clash 2) Your stick to his arm.
Clash 3) Limb to limb - Your hand/arm to his hand/arm.

Statue Drill Studies and Observations 2) The 3 Basic Possibilities

The study is broken down scientifically to the following three possibilities:

Possibility 1) Stick makes contact on arm and stick strikes a worthy target.
Possibility 2) Empty hand makes contact and counter strike with stick.
Possibility 3) Simultaneous hand and stick contact and stick strikes.

Statue Drill Studies and Observations 3) The Statue Arms Vary

Depending on the training assignment, the trainer/statue may:

Trainer Position 1) Have arms straight out and high.
Trainer Position 2) Have arms bent and low.
Trainer Position 3) Have one arm high, the other low.
Trainer Position 4) Have arms pumping.
Trainer Position 5) Steps in and strikes.

Statue Drill Studies and Observations 4) The Protocol Formula

The Drill has 5 trainee body positions:

Trainee Position 1) Outside of the right arm.
Trainee Position 2) Inside of the right arm.
Trainee Position 3) Split between both arms.
Trainee Position 4) Inside the left arm.
Trainee Position 5) Outside the left arm.

Statue Drill 1) Stick Side Contact (strike or block) and Follow-Up Stick Strike
Here we make first contact with the stick (or any weapon. Could be a strike or block. Any contact, and then strike quickly with the stick. The strike could land anywhere appropriate for the circumstances - neck, face, hand, torso or lower. Here, I work across the body, contacting and striking, starting from the outside, then inside, then the obligatory split movement that reminds practitioners to cover against both arms, then inside, then outside.This develops mastery of the stick in your hand.

1) Outside right contact block or strike.

2) Any stick strike.

3) Inside right contact block or strike.

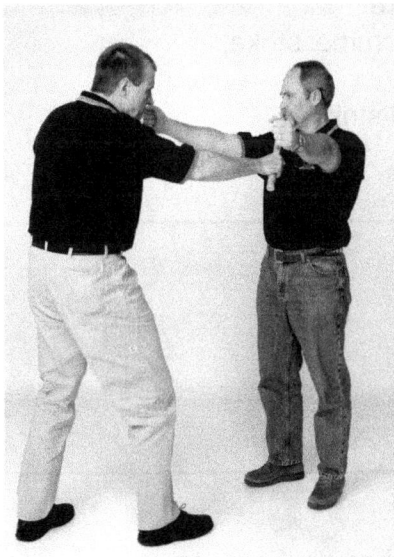

4) Any stick strike.

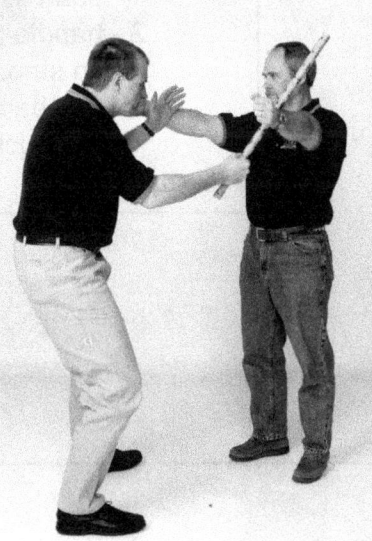

5) Split arm contact.

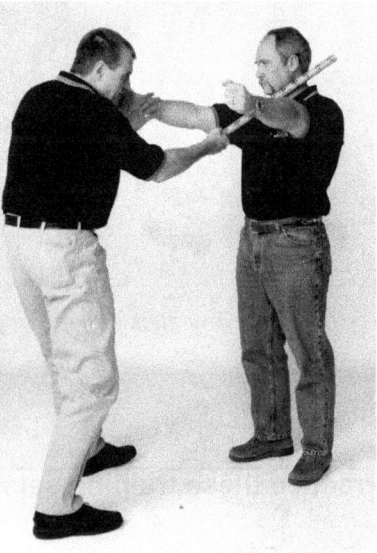

6) Any stick strike.

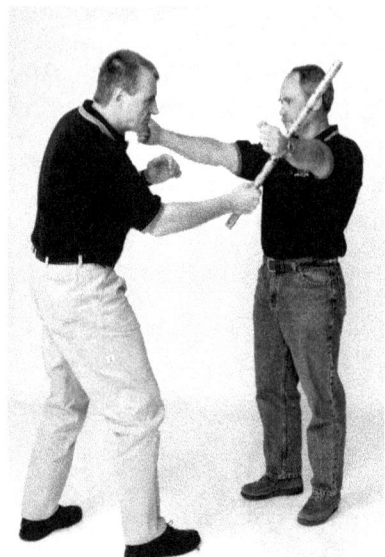
7) Inside left contact strike or block.

8) Any stick strike.

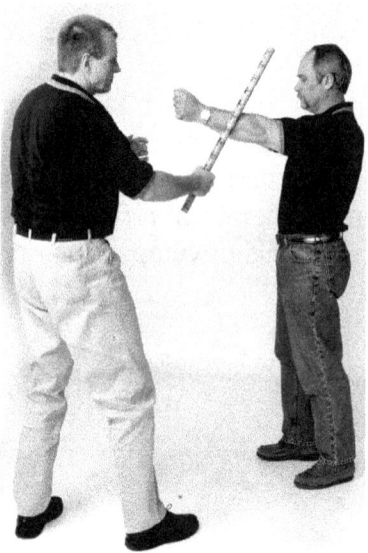
9) Outside left contact strike or block.

Move freely across the front of the trainer. Try to mimic a professional football or basketball player with side-to-side footwork. Don't cross your legs. This is simply an athletic stepping. The strike could come from the shaft, tip, or handle.

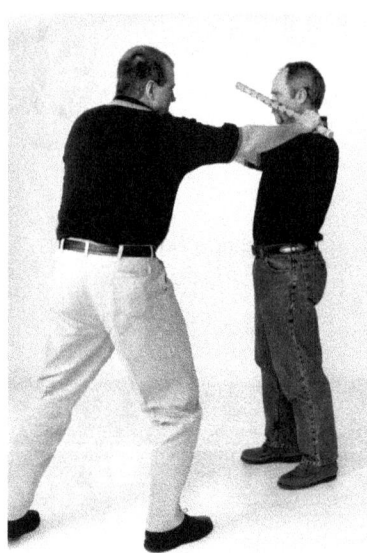
10) Any stick strike.

> You should make a run through with:
> 　　　1 - shaft strike.
> 　　　2 - handle/pommel strike.
> 　　　3 - tip stab.
> 　　　4 - any hand strike.
> 　　　5 - any kick.

Practice these training sets.

Basic Set 1) Stick contact and stick strike - work across the body in five sets.
Basic Set 2) Stick contact and any support hand strike - work across the body in five sets.
Basic Set 3) Stick contact and any kick - work across the body in five sets.

Statue Drill 2) Empty Hand Side Contact and Follow-Up Stick Strike
Here, make contact first with the support hand and/or the arm, and strike quickly with the stick.

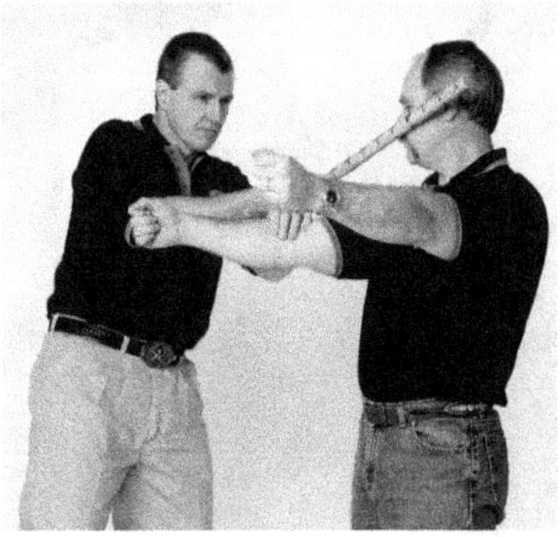

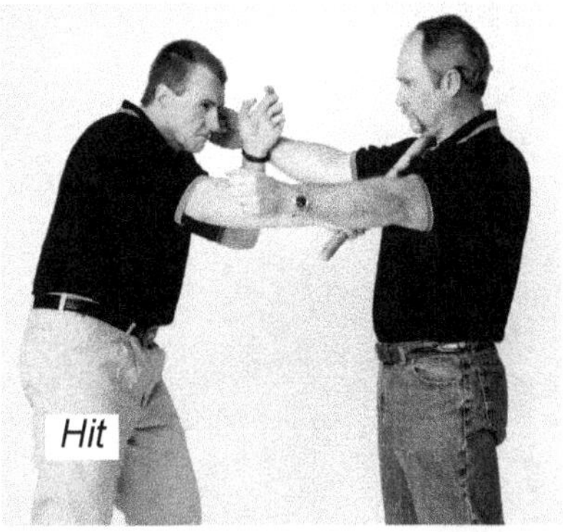

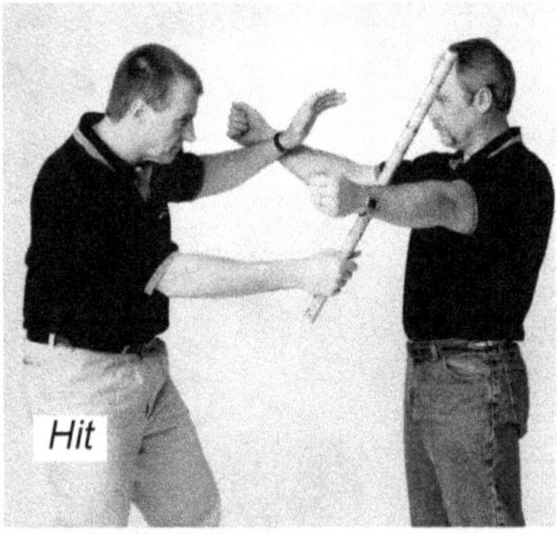

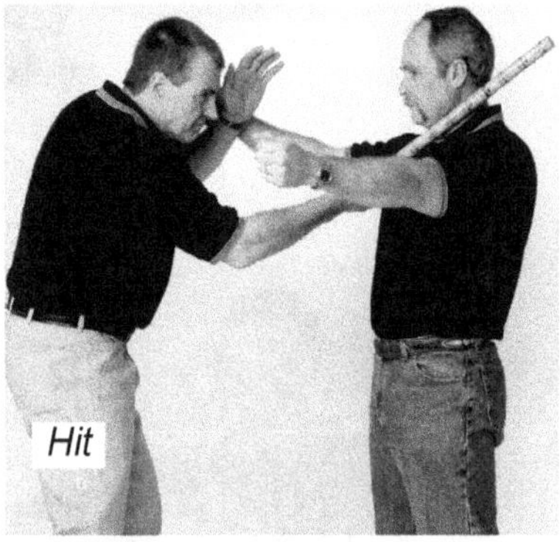

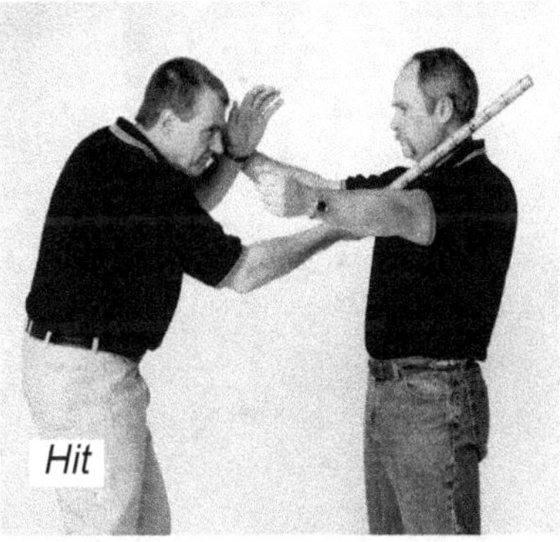

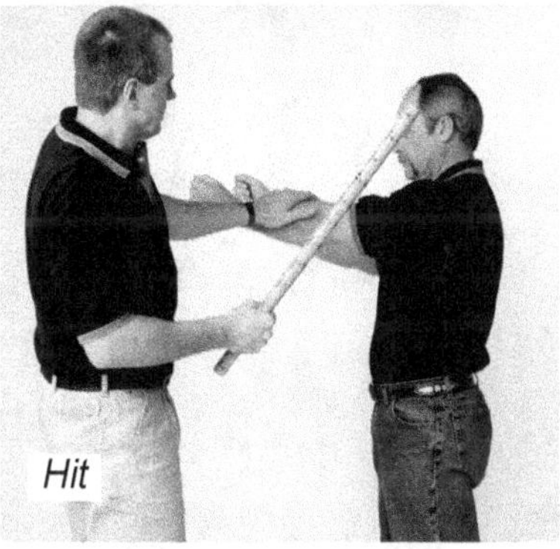

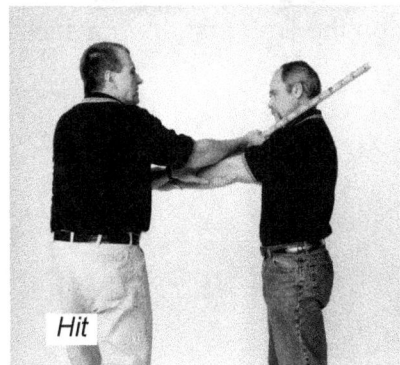

Practice these training sets. One pass across the trainer is a set.

Basic Set 1) Hand contact and stick strike - work across the body in five sets.
Basic Set 2) Hand contact and the same hand strike - work across the body in five sets.
Basic Set 3) Hand contact and kick - work across the body in five sets.

Advanced: Experiment with ground version. The trainer looms over you.

Strike with the shaft, the tip and the handle.

You can "arm" the statue with whatever weapons you need to work with. Knives, guns, whatever.

1 thru 10. 5 contacts and 5 strikes. The trainee works across the trainer using the outside, inside, split, inside, outside formula, making hand contact on the limb first, then a stick strike.

    Experiment with:
- shaft strikes.
- tip strikes.
- handle strikes.
- blocks and grabs.
- standing.
- seated experiments.
- ground experiments.
- arm the statue.

Statue Drill 3) Two-Hand Grip Practice
Next, use a two-handed grip and make a pass across the statue to develop your command and mastery of the two-handed weapon. A great training drill for beginners.

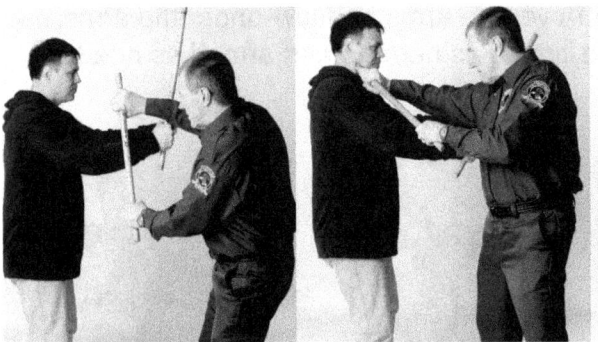

1: Outside arm contact and any hit.　　　2: Inside arm contact and any hit.

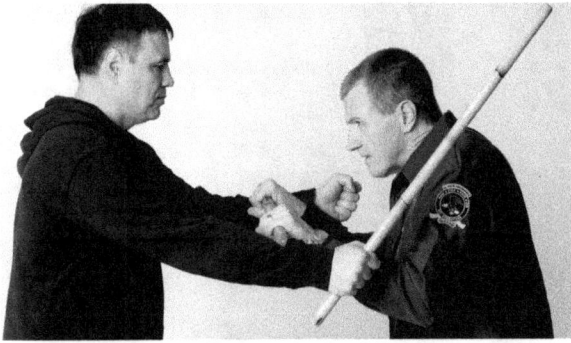
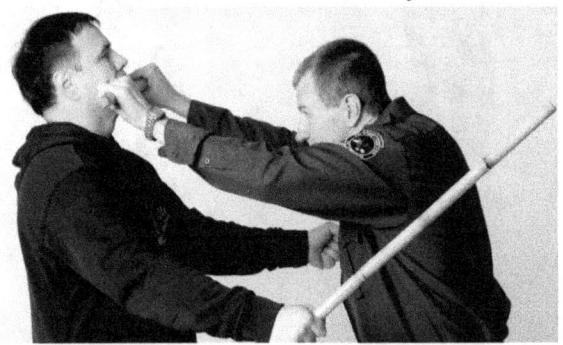

3a: Split arm strike contact, strike arms down.　　　3b: Split arm contact, strike face.

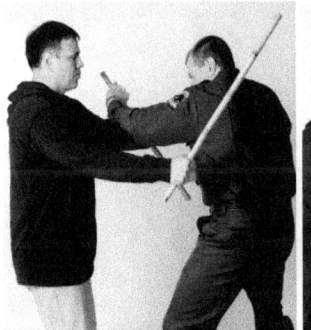
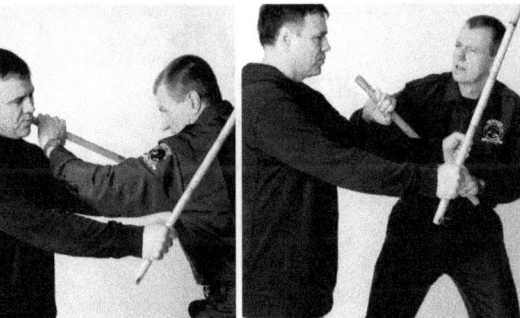

4: Inside other arm contact and any hit.　　　5: Outside other arm contact and any hit.

1 thru 10. 5 contacts and 5 strikes. The trainee works across the trainer using the outside, inside, split, inside, outside formula, making hand contact on the limb first, then a stick strike.

    Experiment with:
- shaft strikes.
- tip strikes.
- handle strikes.
- blocks and grabs.
- standing.
- seated experiments.
- ground.experiments.
- arm the statue.

The Statue - In Summary
The statue drill concept mirrors many martial training practices. It is a great introduction for beginners and great reminder of moves for veterans. It insists on proper coverage of a tactic. Work across the arms and respond high-over the arms and low-under the arms. The statue - your training partner - may be positioned as needed and armed as needed.

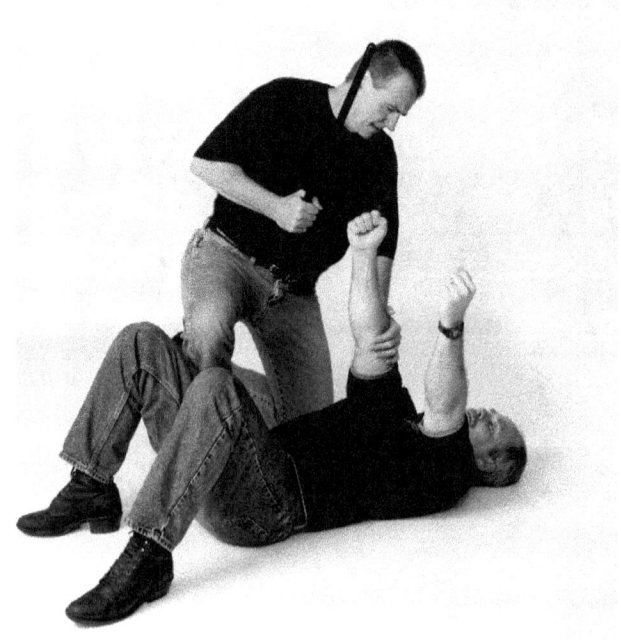

The statue drill can develop grounded/floored perspectives. Trainer or trainee can be topside, bottom-side, both on their right or left sides.

One-Hand and Two-hand Grip Combination Striking - The Vast Arsenal

If you run the numbers on possible combinations of saber and reverse grips, single hand and double hand grips, using shaft, tip and pommel strikes in lunges and hit and retractions, the list would be enormous. Now imagine three strikes and not just two strikes. Now think about four strikes in succession. Now add support strikes and kicks. Your list is long and your arsenal is vast.

Due to this vastness, this section on combination striking will only inspire practitioners to understand and practice a wide variety of strikes. In action, keep all strikes economical, and keep your mind and body free to strike at the best target from the last position you were in with the best weapon you have available.

There is an old, misleading adage, "nearest weapon to the closest target." Actually by definition, the *closest* target may be a somewhat worthless target, or one of no high value. And, if you wait a half-second, or move, you may predict from experience that a better, higher value target may open next. The fight is over sooner with less risk to yourself.

The situation will also dictate what your high value body targets are. If in Afghanistan fighting for your life, your vital targets are different than crowd control duty in Times Square in New York City. As always, fighting is situational, and therefore the *value* of targets is situational.

Use Solo Command and Mastery
Training alone:
- in the air.
- on training objects.
    * wooden posts.
    * rubber tires.
    * heavy bags.
    * body dummies.

Use Partner Training
Working with a trainer/partner as he moves around and holds:
- kicking shields.
- sticks large and small for targets.
- mitts for targets.
- body suits.

For slashing targets. Not to be confused with stick versus stick fighting, the trainer moves and feeds sticks as targets.

The mitt is for stick stabs, support arm strikes and kicks. The stick is for stick slash strikes.

4 Corners and a Brick Wall

A favorite for many, a drill I invented. It opens up the idea of elbow strikes when in a two-handed grip. It is unlikely you will get all 5 strikes, but the drill is supposed to be inspiring. Practice all 5? Maybe get 2, or 3. You position yourself for the drill with your arms and stick off to the left.

>   **Strike 1:** Your right elbow.
>   **Strike 2:** Your stick end.
>   **Strike 3:** The shaft of the stick.
>   **Strike 4:** Your stick end.
>   **Strike 5:** Your left elbow.
>   Repeat but from the right side position.

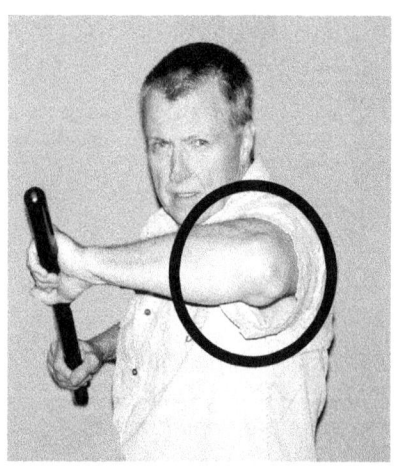

Chapter 11: Impact Weapon Blocking

Sailors participate in a reactionary force, non-lethal, impact weapons class aboard the amphibious command ship USS Blue Ridge.
- Photo by Cynthia Griggs

This book is about blocking, striking and grappling with a stick, and we will cover in this chapter the actual physical steps of study with blocking. Stick-fighting martial arts systems have dozens of stick blocks. There is one art from the Philippines that has 42 blocks, but they split hairs on the angles of the blocks they try to cover.

Often, tactical, self defense and survival programs look to these stick arts like the Filipino programs and simply take all their moves and philosophies and reproduce them, which is an over complicating mistake.

Yes, a researcher or group leader should become adept and knowledgeable of things past, but must have the IQ and wisdom to reduce the abstract and support the who, what, where, when, how and why of their group's needs and mission.

We remind here of our use of the Combat Clock. Basic training is the top, sides and bottom of the clock. 12, 3, 6 and 9. Advanced training covers all 12 numbers of the clock. These numbers are covered through the various hand grips and in standing, kneeling and ground positions. The equation in this study for your solo command and mastery practice is angles, grip and position.

Angles

Basic Training
12 o'clock or top
3 o'clock or your right side
6 o'clock or bottom
9 o'clock or your left side

Advanced Training
1-12 on the clock

+

The Hand Grips

One-handed
 * saber grip
 * reverse grip
 * middle grip

Two-handed
 * one hand on each end
 * two hands on one end

+

Your Position

Standing (moving)

Kneeling

Seated (Get up quick!)

Ground
 * right side and left side
 * top side
 * bottom side

The block you will probably choose will depend completely upon the position your stick was in the very instant before you needed the block. By engraining these blocking movements into your muscle memory you will help facilitate your body's natural reactions to protect yourself with the principle of economy of motion. We will start with the study of unsupported, single-hand blocks.

One Hand Grip, Unsupported Blocking

12 o'clock or high unsupported block.

3 o'clock or right side unsupported block.

6 o'clock or low unsupported block.

9 o'clock or left side unsupported block.

Sometimes you may have to "sweep/slash over and under" the clock.

> **Blocking: One Hand Grip, Basic Training**
> Basic Block 1: 12 o'clock or high.
> Basic Block 2: 3 o'clock or to the right.
> Basic Block 3: 6 o'clock or low.
> Basic Block 4: 9 o'clock or to the left.
> Optional "Sweep under the clock".
> Repeat with the opposite hand.
> Exercise these knee high and on the ground.
>
> **Blocking: One Hand Grip, Advanced**
> Advanced - Block to all 12 numbers on the Combat Clock.
> Repeat with the opposite hand.
> Exercise all these knee high.
> Exercise all these on the ground.

Get the most out of your single hand block, and beware the slashing block

How can you re-enforce the unsupported block? One way to strengthen your unsupported block is to deliver it with force, and leave it in place until the threatening attack is over and the enemy withdraws it to try another attack. We call this the vital concept of "Beware the Slashing Block." If you only slash-block an incoming attack, the attack will continue to invade as you slash past it.

Or you counter attack with your other limb or leg, or you escape with your body. Do not slash at the attack. Beware the slashing attack. Slashing blocks often just nick the incoming attack and, as your block slashes past it, his attack continues inward upon you.

One method to put power into your block is to consider the "hammer the nail" approach. See the photos below for an explanation. This hammering motion forces energy into the shaft or barrel of the weapon, giving you more protection power.

When placing a block, think of hammering a nail with your pommel. This places more force into the shaft and gives you more protection than if you had done this.

One Hand Grip, Supported Blocking - (support hand high, medium or low on the weapon)

12 o'clock or high supported block.

3 o'clock or right side supported block.

6 o'clock or supported block.

9 o'clock or left side supported block.

Blocking: Supported Grip, Basic Training
Basic Block 1: 12 o'clock or high.
Basic Block 2: 3 o'clock or to the right.
Basic Block 3: 6 o'clock or low.
Basic Block 4: 9 o'clock or to the left.
Repeat with the opposite hand.
Exercise these knee high and on the ground.

Blocking: Supported Grip, Advanced
Advanced - Block to all 12 numbers on the Combat Clock.
Repeat with the opposite hand.
Exercise these knee high.
Exercise these on the ground.

The Support Hand Geography

Where on the stick should the support hand or limb go? The support limb should be the hand on stick, forearm on stick, and may even be body parts on stick. A true study of support blocking can go very deeply into a variety of martial arts support blocks that tactical practitioners might shun at first sight. They may well work, but still are shunned. Therefore, in this book, I will only cover very modern and common ones and leave the more questionable ones for other topics and books. These body support blocks can and have worked, but fall under such criticism, they are best left for live seminars and training.

Probably the biggest support mistake I see is when the trainee blocks open-handed palm on stick, with the flat of his hand as displayed on the far left. This is an invitation to a serous hand or finger injury. Your hand's "hammer-edge fist, wrist or forearm is much safer as shown here. There will be reverberations and it will still sting, but it won't be as bad as a smashed open hand.

The forearm or fisted hand may be positioned on the stick near the far end, near the middle, near the hand and, perhaps worst of all from a practical standpoint, right on the top of the hand. In fact, I have seen some of the top tactical proponents place their support palm on the top of their stick hand.

Support high on the stick.   *Support middle of the stick.*   *Support low on the stick.*

The worst and weakest supported block is the support hand atop the stick hand. Most people can easily see why. It offers the least support. Practitioners also have a tendency to keep their hand on this spot, too, from bad repetition training which commits the support and limits counter strikes and other easy, common hand blocks. Part of the confusion is that some stick martial arts often *strike* with the support hand helping to propel the weapon-bearing limb for more speed. This gets confused with blocks and strikes where the power shove is not needed. It also short circuits the mental freedom to easily use the support hand for a host of smart activities. And the worse reason of all to do this is it looks artsy and, when things look cool and artsy, they become addictive mistakes for tactical training.

All these things confuse the practitioner and suddenly, one day, the trainees are over striking and over-blocking with their support hand glued atop their weapon hand. But, this is still a popular block amongst several martial arts stick fighting systems. These systems do not do much hardcore stick combat to actually test that block. As many systems fall prey to the "myth of the duel" as in fighting the mirror image of themselves in stances and strikes, if both fighters do the same thing, it becomes a "Kings X" mistake that washes each other's mistakes out.

The often misused, weakest support block, hand on wrist.

This is the weakest block, collapsing under force.

If you must? The hand must push down on the weapon hand and forward on the stick.

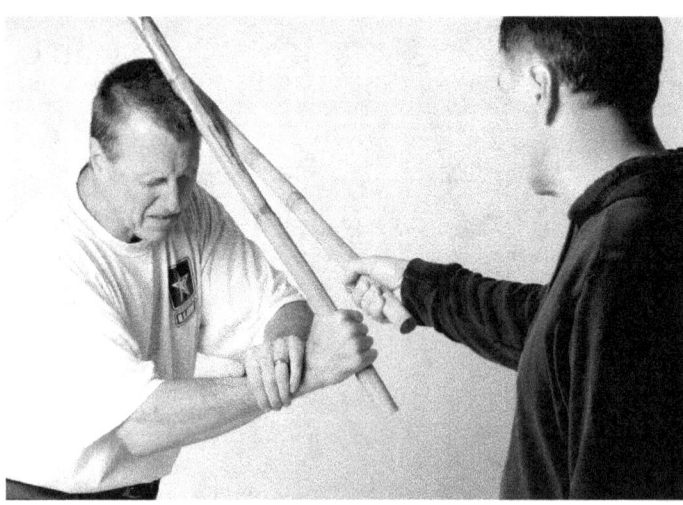
Forget this weak block!

The support limb supports:

- high on stick.
- middle of stick.
- low on stick.

With:
- fist / edge of hand.
- forearm.

Two Hand Grip, Riot/Rifle Grip Blocking

12 o'clock or high block.

3 o'clock or right side block.

6 o'clock or low block.

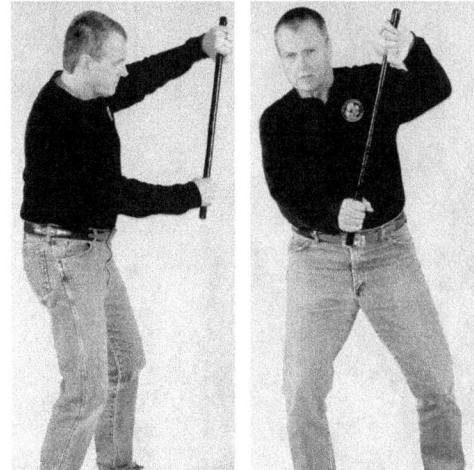
9 o'clock or left side block.

Blocking: 2-Hand Grip, Basic Training
   Basic Block 1: 12 o'clock or high.
   Basic Block 2: 3 o'clock or to the right.
   Basic Block 3: 6 o'clock or low.
   Basic Block 4: 9 o'clock or to the left.
   Optional "Sweep under the clock."
   Repeat with the opposite hand.
   Exercise these knee high.
   Exercise these on the ground.

Blocking: 2-Hand Grip, Advanced
   Advanced - Block to all 12 numbers on the Combat Clock.
   Repeat with the opposite hand.
   Exercise these knee high and on the ground.

Some Two-Hand Grip Blocking Samples versus an Unarmed Attacker Drill
The spread apart, two-hand grip is very successful, popular and very addictive. This is a drill all impact weapon carriers must exercise. The most common attack will be an unarmed one. We call this a mugger exercise as it is quite likely a stick bearer will be attacked by an unarmed thug.

 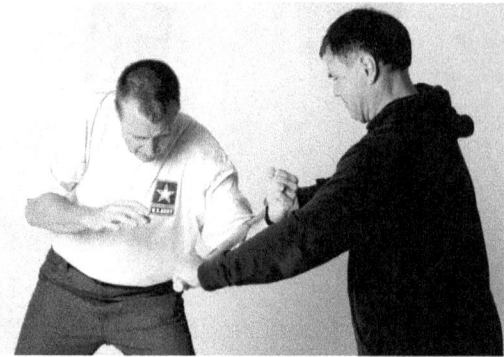

 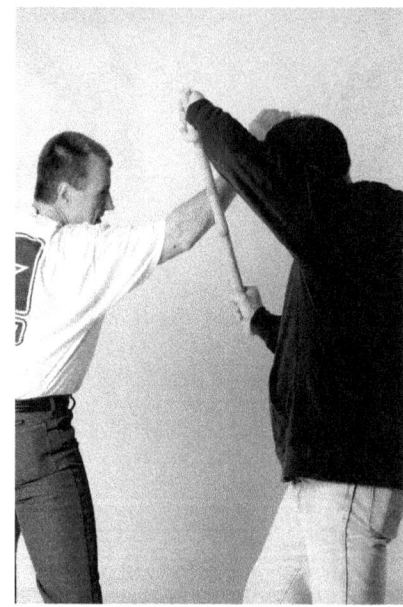 

Two hand grip blocks versus unarmed strikes and kicks
This is a common encounter for officials carrying sticks. This should be practiced by all carriers. An unarmed "mugger" or angry person attacks.

    The trainer attacks with
        - hand strikes
        - kicks
    The trainee blocks with
        - blocks (these can also be powerful strikes)
            * two hands split, both palms down series
            * two hands split, one palm up, one palm down
            * two hands the batting/sword grip.

The Two-Hand Batting or Sword Blocks

Often people grab an impact weapon with both hands on one end, like they are swinging a baseball bat, a cricket bat, hockey stick or sword. It is important that the weapon is long enough for this method to be worthy of such use. Here are the basic and advanced blocking tactics based once again on the Combat Clock.

The 12 o'clock or high block. The hands could be on the right side or the left side.

The 3 o'clock or right side block.

The 6 o'clock or low block. The hands could be on the right side or the left side.

The 9 o'clock or left side block.

Blocking: Two Hand Grip, Batting Basic Training

Basic Block 1: 12 o'clock or high.
Basic Block 2: 3 o'clock or to the right.
Basic Block 3: 6 o'clock or low.
Basic Block 4: 9 o'clock or to the left.
Optional "Sweep over and under the clock."
Repeat with the opposite hand grips.
Exercise these on the ground.

Blocking Two Hand Grip, Batting Advanced Training

Advanced Blocks Set
    Block toward all 12 numbers on the Combat Clock.

Exercise these knee high and on the ground.

Practitioners should realize that the basic 4 and advanced 12 Combat Clock blocking positions will work as a basis for other similar objects, like a walking cane or an empty long gun for just two examples. They stop hooking attacks and pick up and deflect thrusting attacks. Other examples are:

- Police sticks.
- Martial sticks.
- Pipes.
- Tire irons.
- Axe handles.
- Baseball bats.
- Crow bars.
- Etc.

The 12 o'clock or high block.

The 3 o'clock or right side block.

The 6 o'clock or low block.

The 9 o'clock or left side block.

Combat Block and Strike Drills
Now that we have covered strikes and blocks, here are some interactive drills. Do not consider them stick versus stick drills. They are exercises that are meant to develop the trainee's stick skills, and the trainer is holding a stick so that the trainee can hit something very hard.

Work this Basic Series in 6 count sets: Do while moving. These are meant to develop power strikes and competent blocks under stress.

Basic Drill 1: High strike / high block in a set of 6.
Basic Drill 2: Right side strike / side block in a set of 6.
Basic Drill 3: Low stab / low block in a set of 6.
Basic Drill 4: Left side strike / side block in a set of 6.
Basic Drill 5: Mix up the count and angles.
Basic Drill 6: Block 6 punches and kicks.
Basic Drill 7: The port arms or rifle blast and trap strike in sets of 6.
Basic Drill 8: Batting /s word grip versus clock attacks.

Basic Drill 1: The High Strike / High Block in a fast set of 6
In this drill you take turns striking from the 12 o'clock or high position up to 6 strikes. Stop, stalk each other and then exchange strikes again. Make each strike deep and mean. When the attacker strikes realistically hard and deep, defenders regularly use the two-handed grip to stop the real force.

He strikes high on beats 1, 3 and 5.

You strike high on beats 2, 4 and 6.

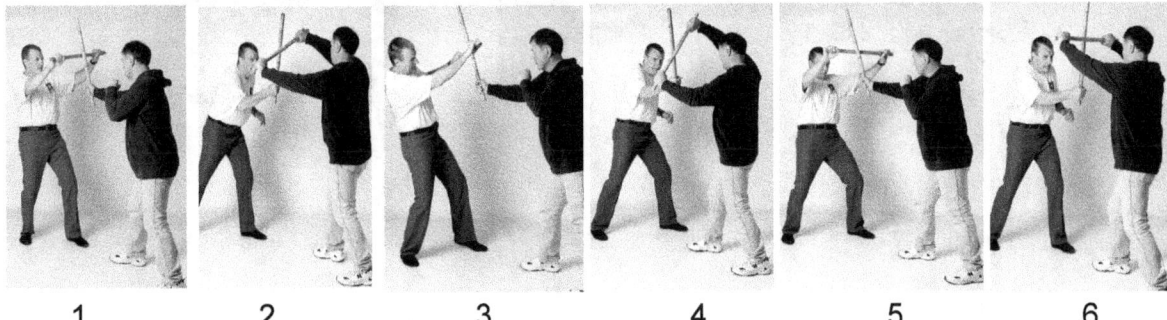

1 2 3 4 5 6

Basic Drill 2: Right side strike / left side block in a fast set of 6

You will strike from your right side or "3 o'clock" side. He blocks. He immediately returns with a power strike. You exchange Combat Clock events.

You strike right on beats 1, 3 and 5.

He strikes back (any side strike) 2, 4 and 6.

Basic Drill 3: Low strike / low block in a fast set of 6

You will strike low or from 6 o'clock side. He blocks. He immediately returns with a low power stab. You exchange 6 blows. Any one-hand or two-handed block is fine to practice.

Any low, or 6 o'clock strike on beats 1, 3 and 5.

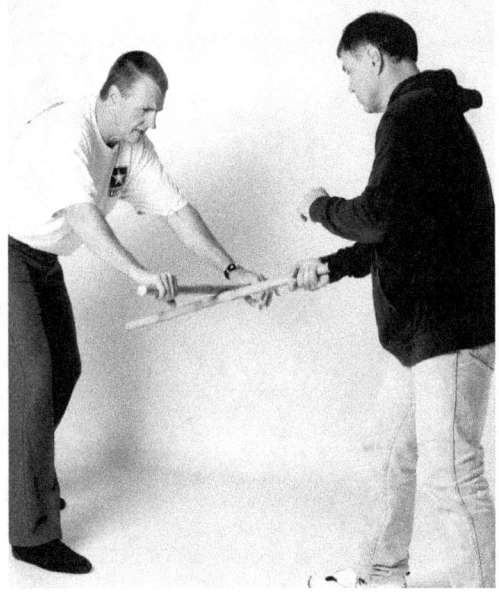

He strikes back on 2, 4 and 6.

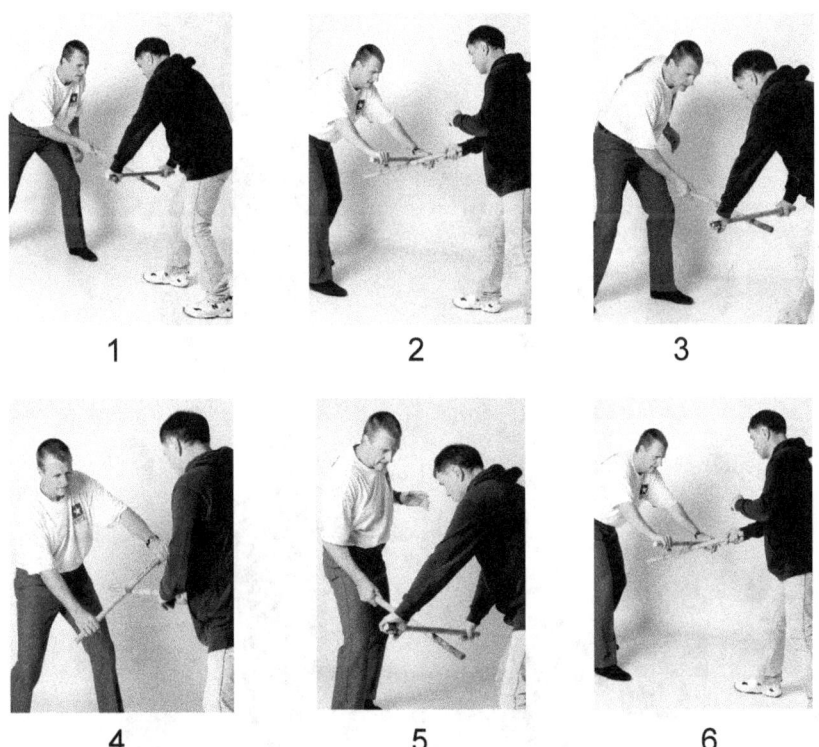

1 2 3

4 5 6

Basic Drill 4: Left side strike and right side block in a fast set of 6
You will strike from your right side or "9 o'clock" side. He blocks. He immediately returns with a power strike. You exchange Combat Clock blows.

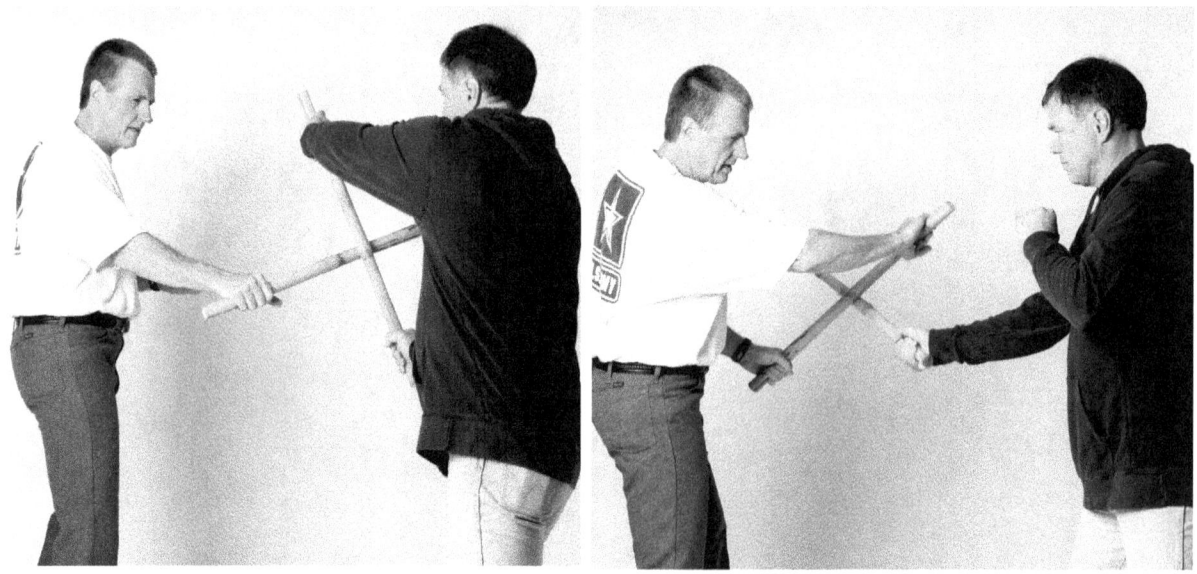

You strike from the 9 o'clock side on beats 1, 3 and 5.   *He strikes on beats 2, 4 and 6.*

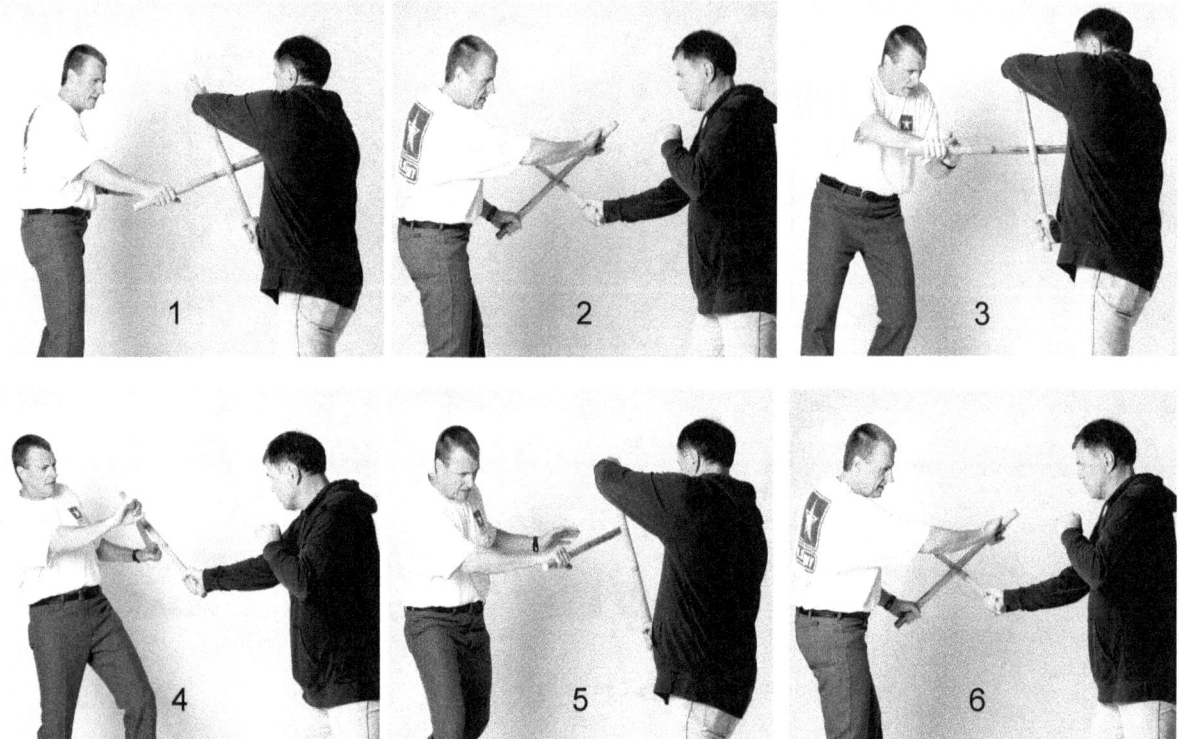

Combination Practice:
    Work the 12 o'clock 6 hits and...
    the 3 o'clock 6 hits and...
    the 6 o'clock 6 hits and...
    the 9 o'clock 6 hits for a total of 24 strikes.

Chapter 12: While Holding. Support Strikes and Kicks

Striking and kicking while holding an impact weapon. A major foundation for close quarters stick combatives is mastering skills in unarmed combatives. Striking and kicking are the subjects of another comprehensive study you must undertake. While training stick combatives, you must also kick and strike "while holding" an impact weapon in support of the stick.

Scientific studies have been conducted proving that footwork and other skills have lessened when athletes held various tools. Overcome this by practicing strikes and kicks "while holding." For example, an athlete cannot run an obstacle course faster when he holds a tool such as a tennis racket. This also relates here to stick-bearing combatives performance. Untrained people punch harder and faster when empty-handed than they do when holding a knife, a stick or a gun in one hand.

Practitioners must train:
1: Single-handed with a saber grip, both right and left-handed.
2: Single-handed with a reverse grip, both right and left-handed.
3: Two hands with a two-handed stick grip.
4: All the above while kicking pads and shields.
5: All practice versus a training partner wearing protective gear.
6: Use the Combat Clock to vary targeting directions and heights.
7: Standing (moving), kneeling and ground practice.

Strike Force Exercise Sets:
Exercise Set 1: Eye attacks/finger strikes while holding a strike.
Exercise Set 2: Palm strikes while holding a stick.
Exercise Set 3: Forearms strikes while holding a stick.
Exercise Set 4: Hammer fists while holding a stick.
Exercise Set 5: Punching while holding a stick.
Exercise Set 6: Elbows while holding a stick.
Exercise Set 7: Body rams while holding a stick.
Exercise Set 8: Limited use of the head butt while holding a stick.
    Limited to avoid knocking yourself out.

Kick Force Exercise Sets:
Exercise Set 1: Front "snap kicks" while holding a stick.
Exercise Set 2: Stomp kicks while holding a stick.
Exercise Set 3: Knee strikes while holding a stick.
Exercise Set 4: Hook kicks while holding a stick.
Exercise Set 5: Side kicks while holding a stick.
Exercise Set 6: Back kicks while holding a stick.

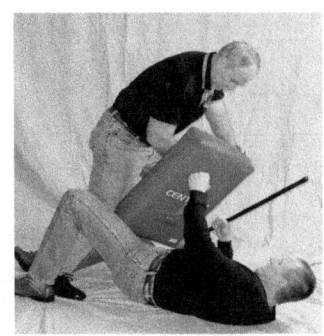

Your "While Holding" Support Strike and Kick Workout

Eye attacks
___ 10 eye attacks while holding a stick, right hand, standing.
___ 10 eye attacks while holding a stick, left hand, standing.
___ 10 eye attacks while holding a stick, right hand, kneeling.
___ 10 eye attacks while holding a stick, left hand, kneeling.
___ 10 eye attacks while holding a stick, right hand, on your back.
___ 10 eye attacks while holding a stick, left hand, on your back.

Palm Strikes
___ 10 palms while holding a stick, right hand, standing.
___ 10 palms while holding a stick, left hand, standing.
___ 10 palms while holding a stick, right hand, kneeling.
___ 10 palms while holding a stick, left hand, kneeling.
___ 10 palms while holding a stick, right hand, on your back.
___ 10 palms while holding a stick, left hand, on your back.

Forearm Strikes
___ 10 forearms while holding a stick, right hand, standing.
___ 10 forearms while holding a stick, left hand, standing.
___ 10 forearms while holding a stick, right hand, kneeling.
___ 10 forearms while holding a stick, left hand, kneeling.
___ 10 forearms while holding a stick, right hand, on your back.
___ 10 forearms while holding a stick, left hand, on your back.

Hammer Fist Strikes
___ 10 hammers while holding a stick, right hand, standing.
___ 10 hammers while holding a stick, left hand, standing.
___ 10 hammers while holding a stick, right hand, kneeling.
___ 10 hammers while holding a stick, left hand, kneeling.
___ 10 hammers while holding a stick, right hand, on your back.
___ 10 hammers while holding a stick, left hand, on your back.

Punching
___ 10 punches while holding a stick, right hand, standing.
___ 10 punches while holding a stick, left hand, standing.
___ 10 punches while holding a stick, right hand, kneeling.
___ 10 punches while holding a stick, left hand, kneeling.
___ 10 punches while holding a stick, right hand, on your back.
___ 10 punches while holding a stick, left hand, on your back.

Elbows
___ 10 elbows while holding a stick, right hand, standing.
___ 10 elbows while holding a stick, left hand, standing.
___ 10 elbows while holding a stick, right hand, kneeling.
___ 10 elbows while holding a stick, left hand, kneeling.
___ 10 elbows while holding a stick, right hand, on your back.
___ 10 elbows while holding a stick, left hand, on your back.

___ 5 kicks while holding a stick, right hand, DMS on your back

Frontal Snap Groin and Shin Kicks
  ___ 5 kicks while holding a stick, right hand, standing.
  ___ 5 kicks while holding a stick, left hand, standing.
  ___ 5 kicks while holding a stick, right hand, on your back.
  ___ 5 kicks while holding a stick, left hand, on your back.
  ___ 5 kicks while holding a stick, two hands standing.
  ___ 5 kicks while holding a stick, two hands on your back.

Stomp Kicks
  ___ 5 kicks while holding a stick, right hand, standing.
  ___ 5 kicks while holding a stick, left hand, standing.
  ___ 5 kicks while holding a stick, right hand, on your back.
  ___ 5 kicks while holding a stick, left hand, on your back.
  ___ 5 kicks while holding a stick, two hands standing.
  ___ 5 kicks while holding a stick, two hands on your back.

Rear Leg Round Kicks
  ___ 5 kicks while holding a stick, right hand, standing.
  ___ 5 kicks while holding a stick, left hand, standing.
  ___ 5 kicks while holding a stick, right hand, on your back.
  ___ 5 kicks while holding a stick, left hand, on your back.
  ___ 5 kicks while holding a stick, two hands standing.
  ___ 5 kicks while holding a stick, two hands on your back.

Front Leg Round Kicks
  ___ 5 kicks while holding a stick, right hand, standing.
  ___ 5 kicks while holding a stick, left hand, standing.
  ___ 5 kicks while holding a stick, right hand, on your back.
  ___ 5 kicks while holding a stick, left hand, on your back.
  ___ 5 kicks while holding a stick, two hands standing.
  ___ 5 kicks while holding a stick, two hands on your back.

Thrust Kicks
  ___ 5 kicks while holding a stick, right hand, standing.
  ___ 5 kicks while holding a stick, left hand, standing.
  ___ 5 kicks while holding a stick, right hand, on your back.
  ___ 5 kicks while holding a stick, left hand, on your back.
  ___ 5 kicks while holding a stick, two hands standing.
  ___ 5 kicks while holding a stick, two hands on your back.

Side Kicks
___ 5 kicks while holding a stick, right hand, standing.
___ 5 kicks while holding a stick, left hand, standing.
___ 5 kicks while holding a stick, right hand, on your back.
___ 5 kicks while holding a stick, left hand, on your back.
___ 5 kicks while holding a stick, two hands standing.
___ 5 kicks while holding a stick, two hands on your back.

Back Kicks
___ 5 kicks while holding a stick, right hand, standing.
___ 5 kicks while holding a stick, left hand, standing.
___ 5 kicks while holding a stick, right hand, on your back.
___ 5 kicks while holding a stick, left hand, on your back.
___ 5 kicks while holding a stick, two hands standing.
___ 5 kicks while holding a stick, two hands on your back.

The Stick Hand Punch

If you train a lot of hardcore, close quarter combatives with sticks, or are an actual veteran of such fights, especially ones on the ground, you are often in too close a range to strike with the tip of the stick, or the shaft and you may even be too close to turn your hand and hit with the pommel. But, you may still be able to punch - with your stick gripping hand. Don't forget this may be an efficient option for standing as well as the ground.

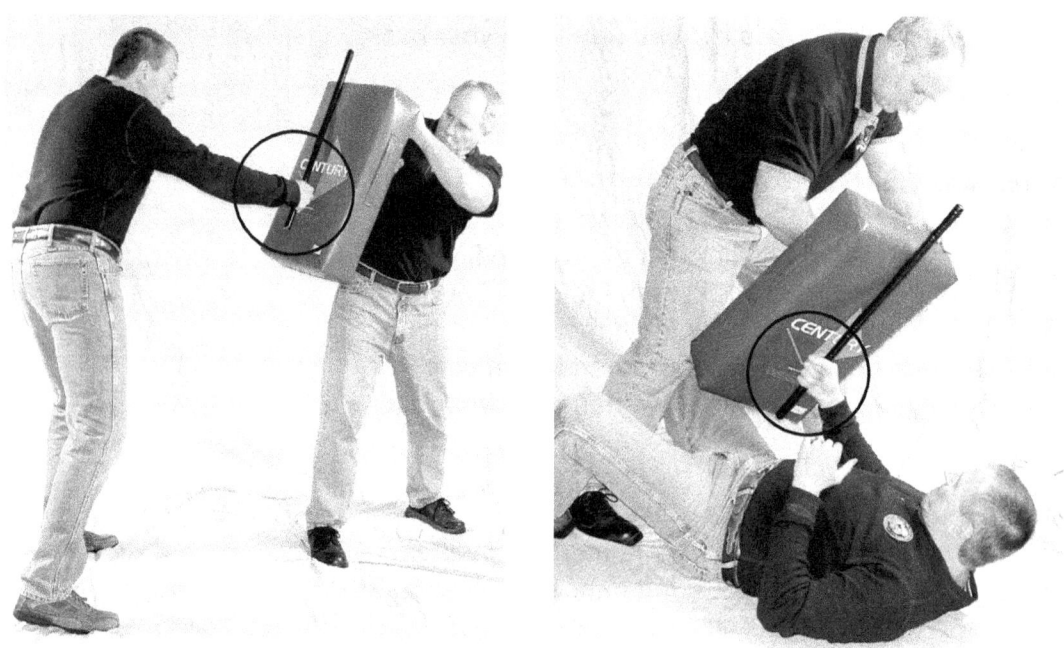

Your hand grip can punch like a fist if need be.

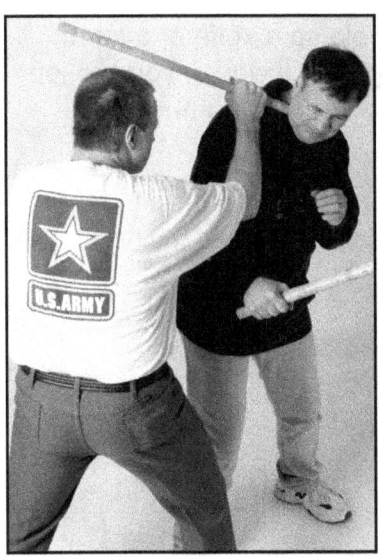

Outside Invasion Series Pommel and/or Shaft Strike

Since we have officially introduced the support hand into the progression, this is a good point to bring in the close-quarters drill set. The action begins with this high backhand reference point. He could strike at you. You could strike at him. Either way, upon contact the reference point has been achieved. You slap/clear his arm. The four step series covers:

Event 1) He doesn't get to block your strike.
Event 2) He barely blocks your strike, with a half-way block.
Event 3) He grabs your incoming wrist.
Event 4) He over blocks your strike.

Event 1: You make forearm to forearm contact, either by striking or blocking, it doesn't matter. In the first series, he fails to block your invasion. You have achieved what Bruce Lee would call a "high reference' point.

You invade by taking your empty hand and slapping hard on his elbow, to clear his arm out of the way. This allows your pommel to invade in and strike to hit the neck or face. Try to keep pressure on his arm. Keep hitting the newly stunned opponent.

The reference point. It matters not who struck or blocked. The connection, the attachment, is made.

We start each series from this point. The free hand pounds the elbow to create an opening.

The weapon side makes a contact, either from striking or blocking.

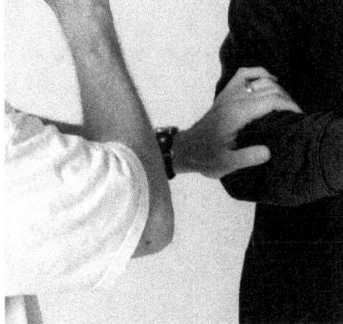

Your support hand could strike the face if in range, but it may have to slap the elbow to clear his arm.

You clear the arm strike and strike the neck.

Remember that these movements will work against a person holding a knife or a lot of other weapons and a person who is unarmed. Also, the strategies will work if you are left-handed versus a left-handed stick man, *but not* with left-handed versus right-handed.

Event 2: The forearm contact is made. In the second series, he center-line blocks your invasion, but only halfway. He barely gets an open hand to the center line of his body. This is not a grab yet, just an open hand, reflexive block. You will:
- twist your torso.
- elbow strike his upper arm and push upon impact.
- clear his arm (a hand grab may be possible).
- continue to hit him.

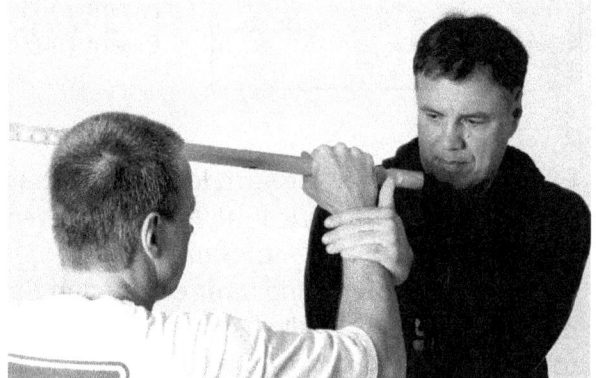

You've cleared a path with your free hand. You strike. He blocks half-way. No grab. Just an open block.

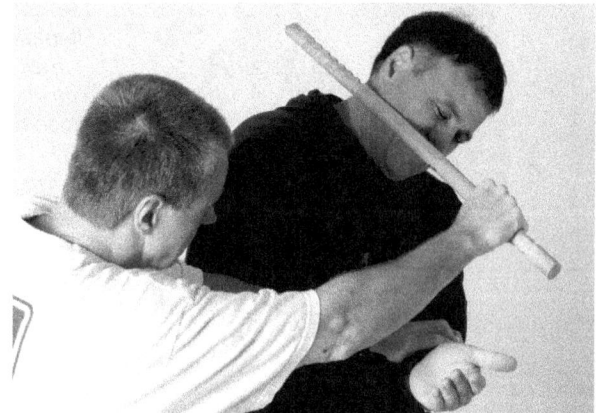

You elbow strike the weapon arm. Drive your body into it. You hand slap the block away.

Again, remember that these movements will work against a person holding a knife or a lot of other weapons and a person who is unarmed. Also, the strategies will work if you are left-handed versus a left-handed stick man, *but not* with left-handed versus right-handed. Switch leads and practice this.

Event 3: You've made forearm to forearm contact. In the third series, he grabs your incoming invasion. There are several options. In this first option, you raise your trapped elbow. You roll that elbow over the top of his arm. Then you elbow strike down on his arm. With your weapon limb now free, you strike until the fight is over. This is the classic elbow rollover release further explained and photographed in an upcoming chapter.

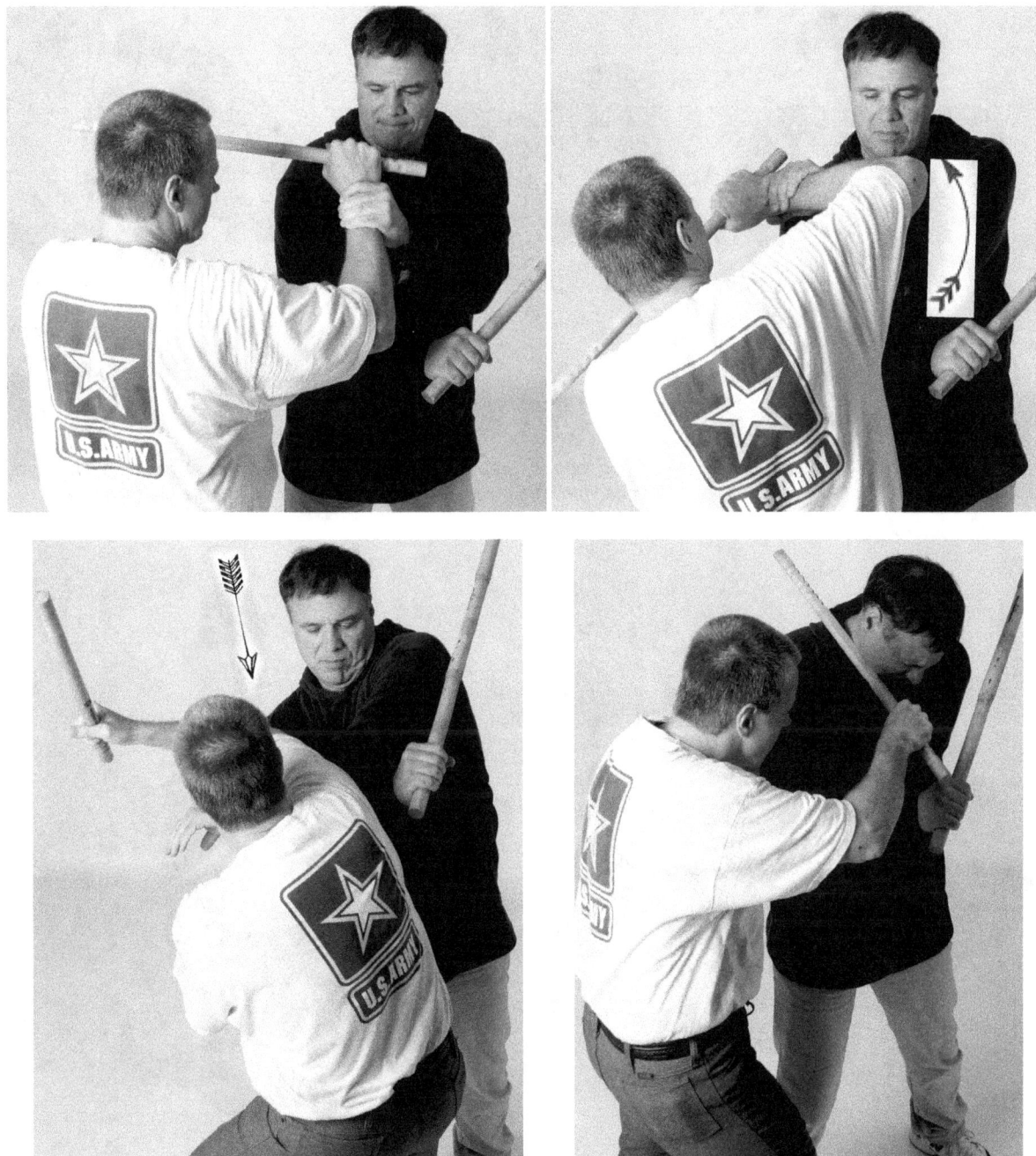

When you roll your elbow up over his arm, it puts his arm in a very weakened position, called by many a *center-lock position*. Then you drive your elbow down over this weakened position, like a downward elbow strike. Once your weapon hand is released from his grip, you can strike with the pommel or the shaft.

In this second option of the third series, you loop under his closest arm. You might turn this looping action into an uppercut punch to the jaw, if possible? Using your knees as a springboard, you might get this jaw strike. Or, you grab the other end of your stick. Get that stick end over his elbow. Try to get the handle of your primary grip over his wrist. Then drop down in a downward spiral and strike until the fight is over.

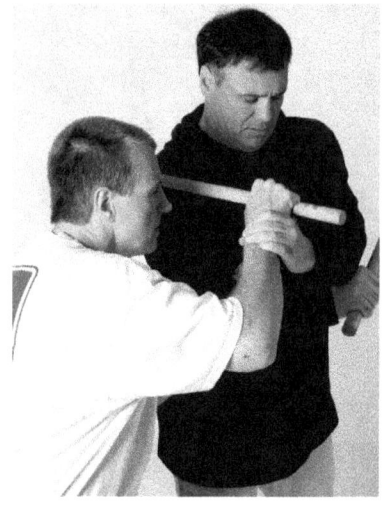

Loop your left hand under his right arm. Punch, if possible.

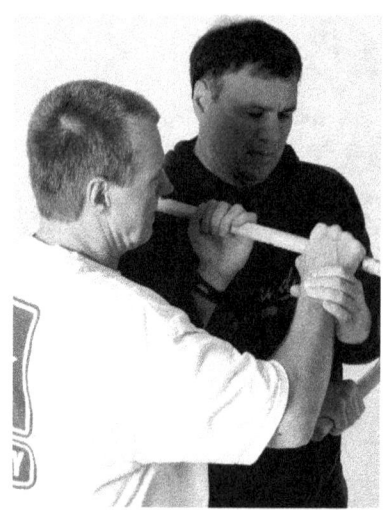

Once looped, grab your stick. If possible, strike with the stick.

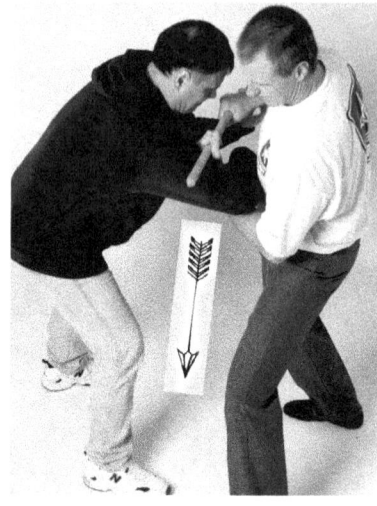

Pull down at an angle. Hook your handle over his wrist.

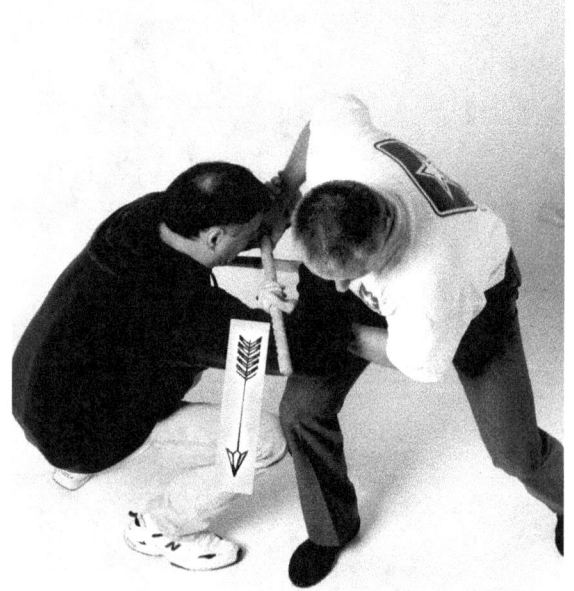

Make sure you hook atop the elbow with the end of the stick.

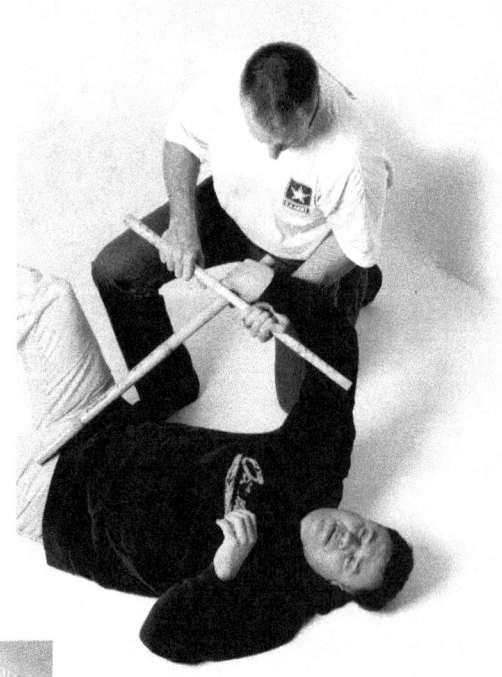

With your left hand under his right, place your stick atop his stick. You may force a disarm by sending the stick downward. Start striking, or choking.

Event 4: You have made forearm to forearm contact In this fourth and last of the basic series, he over-blocks your anticipated attack, a very reflexive response. He passes his center line. Now raise up your first slap contact hand and hook the block, pulling it aside. You strike a critical target. With some practice, you can get very good at this motion. How easy is it? I have seen young children in judo tournaments do this pass and grab move.

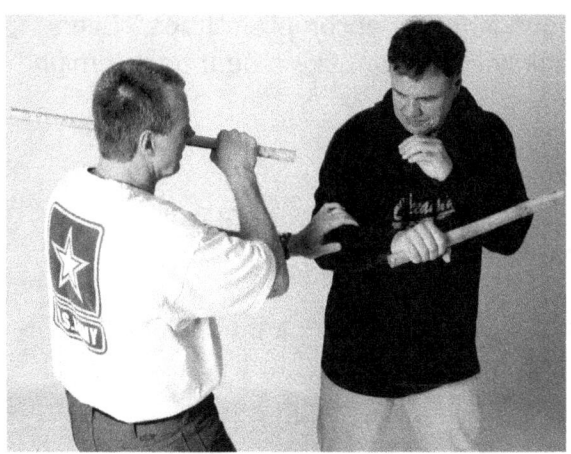 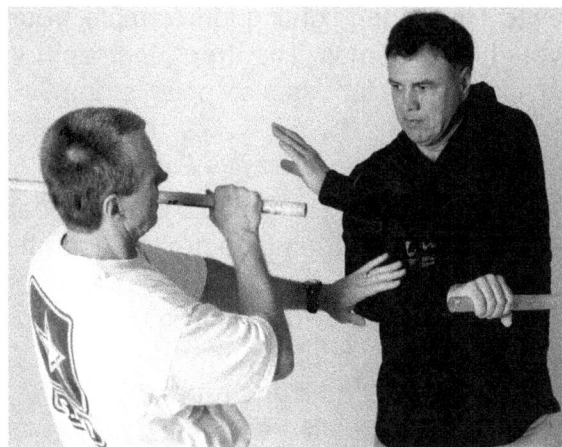

Contact made. The slapping palm comes in on the elbow. You try to pommel strike. Not only does he try to block, he over-blocks and passes his center-line.

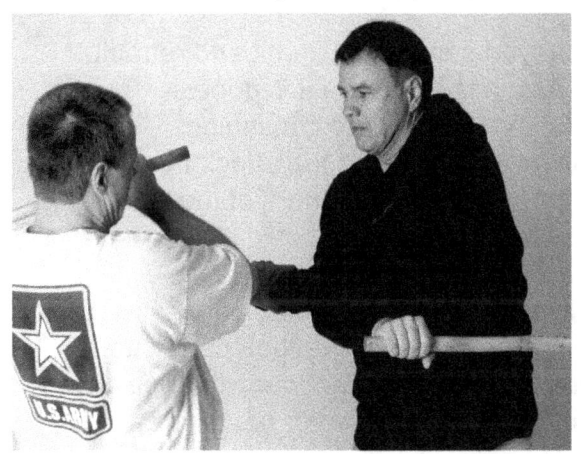 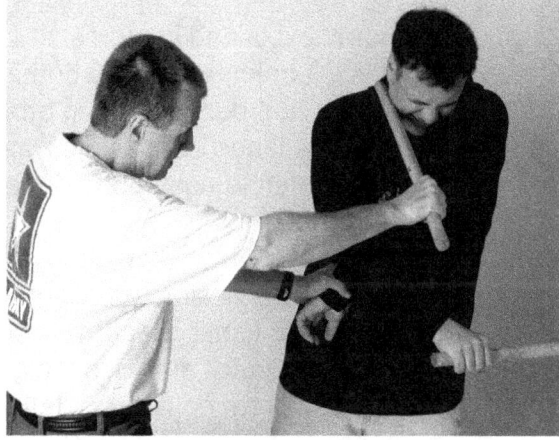

Your slap hand circles up and catches his over-block, pulling his motion over even more. You try to trap his arms in this cross-over and strike.

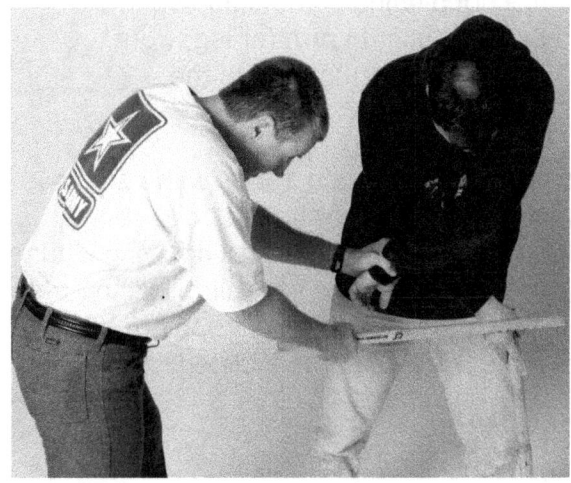

To end all of these sets, we ask that you remember at the end of each to strike the weapon bearing limb to perhaps cause an impact disarm of whatever weapon the trainer is holding, a disarm or some injury that further diminishes his abilities.

I reference here that this is a basic series. It begs the question, is there an advanced series? Yes. There is a ton of close-quarter tricks and plays called by many martial systems as "trapping hands."

The essence of trapping hands is comprised of 4 fundamental movements, I call the 4 Ps. Pinning, passing, pushing or pulling the opponents limbs to clear a path for a significant strike to a significant target. It should be noted that while the term "hands" is used, a person's limbs could also be manipulated by a forearms and weapons applications of the four Ps. And if you use your imagination, in odder times, what other body parts could be used? In ground fighting for example, your torso and legs can accomplish "traps." Even your head at times. The "trap", the capture is certainly not always like a bear trap. It might just last an instant as a stepping stone.

The crash course looks like this:

1: Pinning. Pinning is a push but capturing a limb against his body, a wall, car, whatever and the ground-floor. It is "pinned."

2: Passing: Passing a limb. In the physics of the universe. you are attacked by
- 1: a lunge, or...
- 2: a hit and retract.
- ...delivered either by...
- 3: a thrusting motion, or...
- 4 :a hooking motion.

It is rarely explained in martial studies that you cannot pass a hit and retract. Hand, stick or knife, there is nothing to pass. It is there and it is gone. Retracted. In order to pass an attack you have to have the continuing energy - the lunge - to pass it. Passing involves sweeping the attack from right to left, left to right, up to done, down to up, all the numbers of the Combat Clock.

3: Pulling. Pulling is simply pulling a limb, which involves grabbing, to accomplish a fighting task. You might pull an arm from a weapon carry site and thwart a quick draw. You might pull a limb off a choke. You might pull an arm (or leg) from right to left, left to right, up to done, down to up, all the numbers of the Combat Clock.

4: Pushing. Pushing without pinning. If you are not pinning something you are just pushing it to get it out of the way. You might push an arm (or leg) from right to left, left to right, up to done, down to up, all the numbers of the Combat Clock.

I have been in the "trapping hands," "tapi" business since the late 1980s with various Jeet Kune Do families and Filipino clans. Since I am just trying to train and build savvy, multi-range, multi-weapon, standing-through-ground survivors and not making Wing Chun Kung Fu masters, I just usually present the "trapping" subject in this very simplistic manner Four Ps outline. Pinning, passing, pulling or pushing is the core of all trapping. You can always branch off and obsess on this branch subject if you are so inclined.

Chapter 13: Your Impact Weapon Total Combination Workout list

This is a basic primer for your multiple strikes workout list. Obviously, there are far too many possible combinations to fit in a normal sized book. You must continue to develop the combinations.

Single Slashing Strikes
___ 10 from 12 o'clock or from high and slash back.
___ 10 from 3 o'clock or from the right, and slash back.
___ 10 from 6 o'clock or from low and slash back.
___ 10 from 9 o'clock or from the left and slash back.
    Do with the opposite hand grip.
    Do the above standing.
    Do kneeling.
    Do on the ground.
    Do all with the advanced 12 Combat Clock.

Slashing strikes back and forth on the same, or near same line of attack.

Double Slashing Strikes, somewhat same lines
___ 10 from 12 o'clock or from high and slash back.
___ 10 from 3 o'clock or from the right, and slash back.
___ 10 from 6 o'clock or from low and slash back.
___ 10 from 9 o'clock or from the left and slash back.
    Do with the opposite hand grip.
    Do the above standing.
    Do kneeling.
    Do on the ground.
    Do all with the advanced 12 Combat Clock.

Double Slashing strikes back and forth on the same, or near same line of attack.

Double Slash X Strikes, or Figure Eights (on the clock)
___ 10 from 1:30 or from high.
___ 10 from 10:30 or from the right.
___ 10 from 4:30 or from low.
___ 10 from 7:30 or from the left.
    Do with the opposite hand grip
    Do the above standing.
    Do kneeling.
    Do on the ground.
    Do all with the advanced 12 Combat Clock.

The X: Slashing strikes back and forth not on the same line of attack.

Triple Slashing Strikes, somewhat same lines
___ 10 from high, slash back and back.
___ 10 from the right, slash back and back.
___ 10 from the low, slash back and back.
___ 10 from the left, slash back and back.
    Do with the opposite hand grip.
    Do the above standing.
    Do kneeling.
    Do on the ground.
    Do all with the advanced 12 Combat Clock.

Triple slashing strikes back and forth on the same, or very near same line of attack.

Fanning Strikes
___ 10 from 12 o'clock or from high, fan to low.
___ 10 from 3 o'clock or from the right, fan to left.
___ 10 from 6 o'clock or from low, fan to high.
___ 10 from 9 o'clock or from the left, fan to right.
    Do with the opposite hand grip.
    Do the above standing.
    Do kneeling.
    Do on the ground.
    Do all with the advanced 12 Combat Clock.

The fanning strike.

Circular Strikes
___ 10 from 12 o'clock or from high for a complete oval.
___ 10 from 3 o'clock or from the right for complete oval.
___ 10 from 6 o'clock or from low, for a complete oval.
___ 10 from 9 o'clock or from the left for a complete oval.
    Do with the opposite hand grip.
    Do the above standing.
    Do kneeling.
    Do on the ground.
    Do all with the advanced 12 Combat Clock.

Big circles and small circles.

Slashes and Stabs
___ 10 from 12 o'clock or from high slash, any tip stab.
___ 10 from 3 o'clock or from the right slash, any tip stab.
___ 10 from 6 o'clock or from low slash, any tip stab.
___ 10 from 9 o'clock or from the left slash, any tip stab.
    Do with the opposite hand grip.
    Do the above standing.
    Do kneeling.
    Do on the ground.
    Do all with the advanced 12 Combat Clock.

Slash and then stab.

Stabs and Slashes
___ 10 from 12 o'clock or from high tip stab, any slash.
___ 10 from 3 o'clock or from the right tip stab, any slash.
___ 10 from 6 o'clock or from low tip stab, any slash.
___ 10 from 9 o'clock or from the left stab, any slash.
    Do with the opposite hand grip.
    Do the above standing.
    Do kneeling.
    Do on the ground.
    Do all with the advanced 12 Combat Clock.

Stab and then stash.

Single-Hand to Double-Hand Strikes
___ 10 from 12 o'clock or from high, single to double.
___ 10 from 3 o'clock or from the right, single to double.
___ 10 from 6 o'clock or from low, single to double.
___ 10 from 9 o'clock or from the left, single to double.
    Do with the opposite hand grip.
    Do the above standing.
    Do kneeling.
    Do on the ground.
    Do all with the advanced 12 Combat Clock.

Single hand to double hand.

Double-Hand to Single-Hand Strikes
___ 10 from 12 o'clock or from high, double to single.
___ 10 from 3 o'clock or from the right, double to single.
___ 10 from 6 o'clock or from low, double to single.
___ 10 from 9 o'clock or from the left, double to single.
    Do with the opposite hand grip.
    Do the above standing.
    Do kneeling.
    Do on the ground.
    Do all with the advanced 12 Combat Clock.

Double hand to single hand.

Support Hand and Then Weapon Strikes
___ 10 empty hand strikes followed with any stick strike.
___ 10 stick strikes followed with any empty hand strike.
    Hand strike and then stick strike.
    Stick strike, then hand strike.
    Do with the opposite hand grip.
    Do the above standing.
    Do kneeling.
    Do on the ground.
    Do all with the advanced 12 Combat Clock.

Hand then stick. Stick then hand.

Support Hand and Then Weapon Strikes
___ 10 empty hand strikes followed with any stick strike.
___ 10 stick strikes followed with any empty hand strike.
    Do with the opposite hand grip.
    Do the above standing.
    Do kneeling.
    Do on the ground.
    Do all with the advanced 12 Combat Clock.

Empty hand strike and then any stick strike.

Support Kicks and Weapon Strikes
___ 10 kicks followed with any stick strike.
___ 10 stick strikes followed with any kick.
    Do with the opposite hand grip.
    Do the above standing.
    Do kneeling.
    Do on the ground.
    Do all with the advanced 12 Combat Clock.

Any support kicks, then any empty hand strike.

Stick, Hand, Kick in Combinations
___ 10 hand strikes, kicks, stick strikes.
___ 10 kicks, hand strikes, stick strikes.
___ 10 stick strikes, hand strikes, kicks.
___ Continue the variations.
    Do with the opposite hand grip.
    Do the above standing.
    Do kneeling.
    Do on the ground.
    Do all with the advanced 12 Combat Clock.

Continue to develop more combinations.

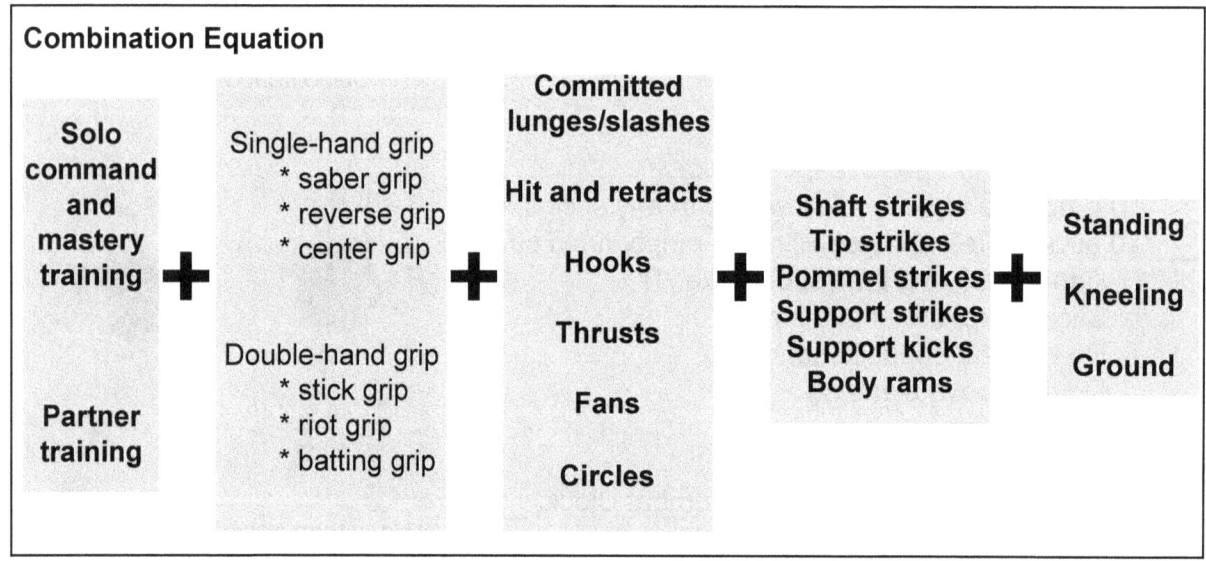

Chapter 14: Impact Weapon Retention

What happens when he grabs your impact weapon on your belt? When you reach for it? He wants it when you pull it. He wants it when you threaten with it. He wants it when you use it.

The Grab Continuum

Your impact weapon may be removed during this continuum from carry site to use:
   1: Removed from its carry site.
   2: Removed from you while attempting to first touch and first draw your weapon.
   3: Removed from you while you are drawing the weapon.
   4: Removed from you while you present the weapon.
   5: Removed from you while you are using the weapon in strikes and blocks.

Attacks on Sheaths and Lanyards and Carry Sites

Retention begins at the carry site. You can purchase a carrier that will limit the removal of your baton/stick, but it will never fully ensure that an opponent won't strip a stick from your belt line or pockets or carry site.

A breakaway sheath sample. The closed baton breaks out the opening, either by your hands, or your enemy's hands.

A very common baton, carrier loop. The baton has a rubber grommet and slips down into the ring, stopping at the grommet. Any opponent can pull this stick out also.

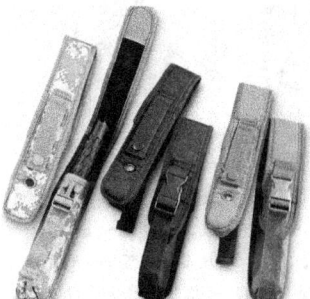

There are many versions of sheath carries. None really offer any significant retention capabilities.

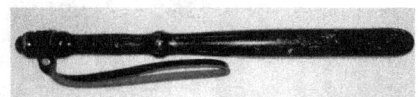

Many new and old batons have retention straps/lanyards that a person can wrap around their hand or wrist.

Problem Set: The Weapon Grabs on Your Carry Site

The best way to start this study is to identify the problems right from your carry site. There are only so many ways an opponent can grab at your weapon from your carry site. It would be an overload to photograph each attack from every possible height and number of the Combat Clock. The basic hand grabs at your carry site are:

Grab 1: His left hand - "same-side grab," he grabs at your right side weapon with his left hand.
Grab 2: His right hand - "cross-grab," he grabs at your right side carry with his right hand.
Grab 3: His left hand - "cross grab," he grabs at your left side weapon with his left hand.
Grab 4: His right hand - "same side grab," he grabs at your left side weapon with his right hand.
Grab 5: Third party grabs at the right side carry site.
Grab 6: Third party grabs at the left side carry site.
Grab 7: One or both hands grab from behind.
Grab 8: On Ground: Applications of the above.
* He is topside making these grabs.
* He is bottom-side making these grabs.
* Some are possible side-by-side.

Grab 1: His left hand - "same-side," he grabs at your right side weapon with his left hand.

Grab 2: His right hand - "cross grab," he grabs at your right side carry with his right hand.

Grab 3: His left hand - "cross grab," he grabs at your left side weapon with his left hand.

Grab 4: His right hand - "same side grab" he grabs at your left side with his right hand.

Grab 5: Third party, right side attack.

Grab 6: Third party, left side attack.

Grab 7: One or both hands grab from behind.

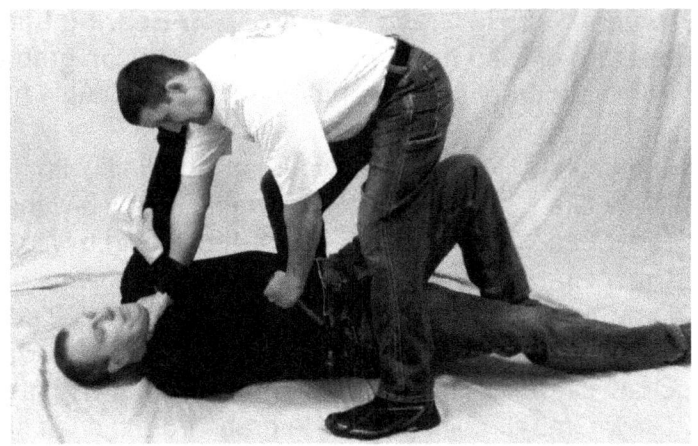

Grab 8: Worry about grabs on the ground while you are topside, bottom-side or side-by-side.

Problem Set: Attacks with Basic Big Arm Wraps Interrupting Your Quick Draw

It is common for the opponent to snatch your impact weapon while it is out, open and either being presented for a threat or used. These big arm catches can be completely reflexive moves.

Arm Wrap 1: Right arm wrap - a common, reflexive close quarters weapon catch.
Arm Wrap 2: Left arm wrap - a common, reflexive close quarters weapon catch.

Right arm wrap - a common, reflexive close quarters weapon catch.

Double arm wraps of your weapon and weapon bearing limb.

Problem Set: He Grabs Your Weapon-Bearing Limb As You Draw

There are only so many ways an opponent can grab your weapon limb as you try to draw it from your carry site. The basic single and double hand grabs at your carry site are:

   Draw Grab 1: He grabs at your left side limb with his right hand.
   Draw Grab 2: He grabs at your right side limb with his right hand.
   Draw Grab 3: He grabs at your right side limb with his left hand.
...Draw Grab 4: He grabs at your left side limb with his left hand.
...Draw Grab 5: Both hands - He grabs at your left side weapon carry site with both hands.
...Draw Grab 6: Both hands - He grabs at you right side weapon carry site with both hands.
...Draw Grab 7: Third party - He grabs from the right side while you are preoccupied.
...Draw Grab 8: Third party - He grabs from the left side while you are preoccupied.
...Draw Grab 9: One or both hands grab from behind.
   Draw Grab 10: Ground - all grabs on ground - applications of the above:
       * He is topside or bottom-side, or possible side-by-side making grabs.

Grab 1: His right hand - "same-side grab," He grabs at your left side weapon limb with his right hand.

Grab 2: His right hand - "cross-grab," He grabs at your right side limb with his right hand.

Grab 3: His left hand - "same-side grab," He grabs at your right side weapon limb with his left hand.

Grab 4: His left hand - "cross-grab," He grabs at your left side limb with his left hand.

Grab 5: Both hands - He grabs at your right side weapon limb with both hands.

Grab 6: Both hands - He grabs at your left side weapon limb with both hands.

Grabs 7 and 8: Third party. He grabs your weapon bearing limb from the right or left side while your attention is pre-occupied with people in front of you.

Grab 9: He grabs your weapon limb from behind you.

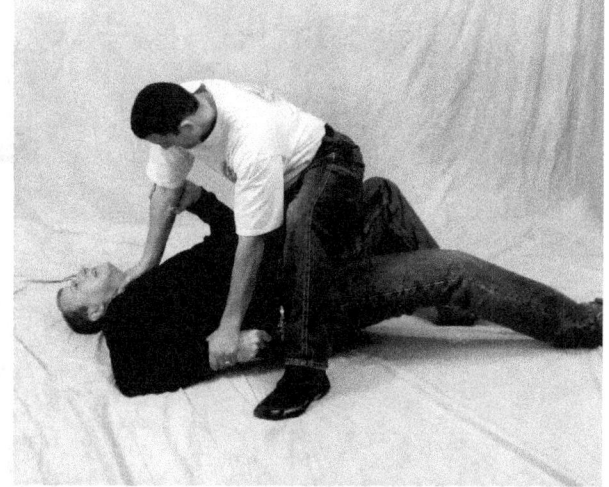
Grab 10: He grabs your forearm on the ground. opside, bottom-side or side-by-side.

Problem Set: Double-Hand Grabs onto the Weapon

It is common for the opponent to snatch your impact weapon wether it is out, open and, either when it is in use, or being presented for a threat. These double-hand catches can be completely reflexive moves on the part of the opponent and, usually from the front. If from the side, the solutions for this are usually the same as the frontal grabs.

High Grab Series
    Double-Hand Grab 1: High grab on your right-hand stick. Standing or ground.
    Double-Hand Grab 2: High grab on your left-hand stick. Standing or ground.

Low Grab Series
    Double-Hand Grab 1: Low grab on your right-hand stick. Standing or ground.
    Double-Hand Grab 2: Low grab on your left-hand stick. Standing or ground.

Your stick is high, and he double hand grabs on it - catching your right-handed stick in the first photo and your left-handed stick in the second photo.

Your stick is low, and he double hand grabs on it - catching your right-handed stick in the first photo and your left-handed stick in the second photo.

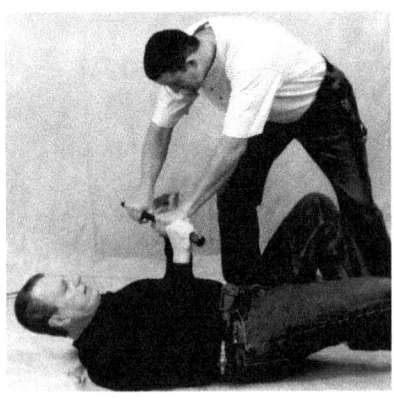

Any two handed grabs while you are on the ground in any ground position.

On the right, here is a sample of a split grab. One hand grabs the weapon limb, the other grabs your support hand.

Retention Problem Set: He Grabs Your Drawn Stick with a Single Hand

There are only so many ways an opponent can grab your weapon with a single hand.

Drawn Grab 1: He grabs your left-handed, held high stick with his right hand.
Drawn Grab 2: He grabs your right-handed, held high stick with his right hand.
Drawn Grab 3: He grabs your right-handed, high stick with his right hand.
Drawn Grab 4: He grabs your left-handed, high stick with his left hand.
Drawn Grab 5: He grabs your right-handed, low stick with his left hand.
Drawn Grab 6: He grabs your right-handed, low stick with his left hand.
Drawn Grab 7: He grabs your left-handed, low stick with his right hand.
Drawn Grab 8: He grabs your left-handed, low stick with his left hand.
Drawn Grab 9: Third party - He grabs from the right side while you are preoccupied.
Drawn Grab 10: Third party - He grabs from the left side while you are preoccupied.
Drawn Grab 11: Ground - All grabs on ground, high and low applications of the above.
        * He is topside or bottom-side, or possible side-by-side making grabs.

Grab 1: His right hand - "cross-side grab," He grabs at your left side weapon limb with his right hand.

Grab 2: His right hand - "same-grab," He grabs at your right side limb with his right hand.

Grab 3: His left hand - "same-side grab," He grabs at your right side weapon limb with his left hand.

Grab 4: His left hand - "cross-grab." He grabs at your left side limb with his left hand.

Grab 5: Left grab to right hand stick. Same side grab. Low stick.

Grab 6: Cross grab. Right grab to right hand stick. Low stick.

Grab 7: Right grab to left hand stick. Same side grab. Low stick.

Grab 8: Left grab to left hand stick. Cross grab. Low stick.

Grab 9 and 10: He grabs your weapon limb from behind you.

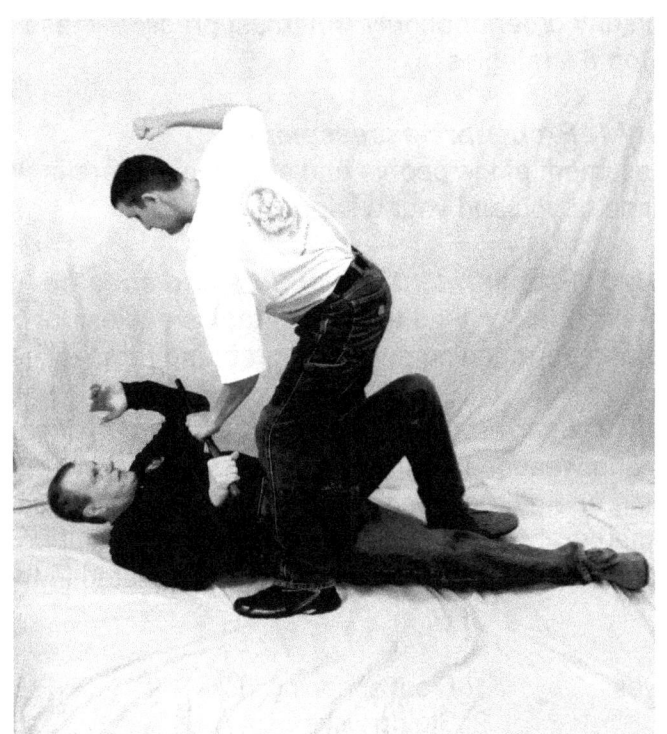

Grab 11: He grabs your weapon in the ground fighting matrix.

Basic Retentions Solutions

These are not listed in any order or priority. You must practice these retention methods in the *Stop 1* through *Stop 6* strategies.

Solution Possibility 1: Proper pre-assessment
Make a proper assessment of the people and situation you are in. Will they fight you? Are they in charging range of you and your weapon?

Solution Possibility 2: Proper retentive holsters and sheaths
This solution is self-explanatory. Find the carrier that is easiest for you to draw and at the same time hardest for a surrounding enemy to grab and remove.

Solution Possibility 3: Learn the core combatives hand strikes, blocks and kicks.
All weapon fighting and weapon retention is based on a working knowledge of unarmed combatives. When the enemy reaches to grab your weapon, you can strike out and block the incoming attempted grab. You strike and kick him before, during and after a grab. Learn these and practice these with regularity. These are listed in the "While Holding" chapter.

1: Finger attacks to eyes
2: Palm strikes
3: Forearm strikes
4: Hammer fists
5: Punching
6: Elbow strikes
7: Body rams
8: Limited use of the head butt

1: Front snapping kicks
2: Stomp kicks
3: Knees
4: Round kicks
5: Side kicks
6: Thrust kicks
7: Back kicks

1: Block high (12 o'clock)
2: Block right (3 o'clock)
3: Block low (6 o'clock)
4. Block left (9 o'clock)

Solution Possibility 4: Secure the weapon
Having solutions 1 and 2 in place, secure the weapon with your hand if the weapon is grabbed at the carry site. Secure the weapon with one hand, if the weapon itself is grabbed on belt/carry site. You may also need to grab and freeze his grabbing limb. Once you know the weapon is safe, torque your body - the weapon-side - violently away as you strike and/or push the adversary, using strikes, body rams and kicks.

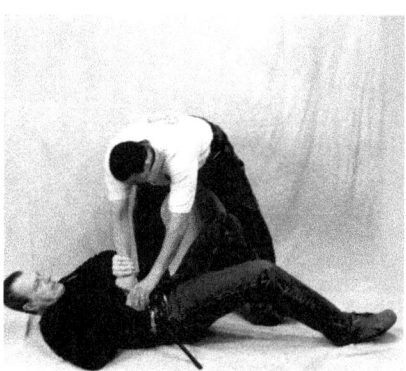

Same side or cross-grab, first secure the weapon by grabbing the arm as a first step. Standing or ground.

Solution Possibility 5: Release and step-away

If the adversary grabs your weapon bearing limb instead of the weapon itself as you attempt to draw your weapon (pistol, knife or stick), he is not actually grabbing your weapon! This grab will often cause professionals to rage into "weapon-retention-mode frenzy," tying up the other hand with a retentive grab, when it could be free for fighting. Instead, let go of your weapon, take your captured hand from the weapon and step back away. A slap release or circular release, or any other release will free your limb.

He grabs your wrist or forearm, NOT THE WEAPON, as YOU grab your weapon. You let go of your grip on your weapon and bring your hand forward and away from the weapon. Step back.

Step back with your weapon side and slap release his grip, or use any number of releasing methods to free for limb. Try to draw again, if needed, or your next best choice.

Always watch out for the other hand. He may also attempt a strike or an additional grab. Use body evasion and blocking skills.

Solution Possibility 6: Basic Releasing Techniques

Release 1: Pull Aways

If your limb is grabbed, you may gain a release by a powerful pullback in an optimum direction. That optimum direction usually involves ripping through the opening of the gripping hand.

With powerful motion involving the body, yank your arm free. The best direction usually involves using the opening in his grip. This works versus all kinds of grabs in various angles and positions.

Release 2: Circular Releases

If your arm is grabbed, you may gain a release by turning/twisting your hand around the grabber's hand in a clockwise or counter-clockwise manner.

Whether it's your empty hand or weapon hand, circle the grabbing hand/wrist/forearm.

Release 3: The Slap Release, Push-Pulls

If your arm is grabbed, you may gain a release by slapping at the gripping forearm.

Whether empty hand or weapon hand, "slap" while using a push-pull force.

Release 4: The Elbow Roll-Overs

This move involves getting the attacker into a painful and weakened position that martial practitioners call the "S" lock, "V" lock, "Z" lock or "center lock." In this position, it is difficult for someone to maintain grip strength. This may resemble a circular release.

This S, V, Z or center lock. It is as though he is driving with his right hand and he makes a left turn.

This S, V, Z or center lock. It is as though he is driving with his left hand and he makes a right turn.

If your forearm is grabbed, you may gain a release by lifting your elbow up and out high. Next, you move as though you are going to elbow strike the face of the attacker. You probably will not reach him, but this committed motion should twist his gripping arm into the center lock position.

Next, you "elbow-strike" downward over his arm. Drop your body with knee bends also and in unison. This will most likely garner a release. Elbow up. Elbow over. Elbow down.

He grabs. You raise your elbow over his grab. You roll your elbow over the top of his arm.

Elbow strike downward. Bend at the knees with this movement to assist the force. This will usually cause a release.

Continue to fight on as needed.

Release 5: Rowing
Rowing usually also uses the concept of the "center lock." Remember your strikes and kicks.

Rowing 1: Row through a one-hand grab
If the attacker grabs your impact weapon with one hand, you double grab your stick. Try a quick pull first. You yank the weapon out, or at worst, take away his balance. Lift the outside tip of your stick, and you can see that the attack is already turning into the center lock weakened position. Drive the heightened tip down toward his center line as you forearm strike his forearm.
The combination of the driving stick and the forearm smash should release the stick from his grip.

He grabs. You grab your stick with two hands. Outside tip up. Tip over. Tip down. Hit forearms.

Don't forget that when he has two hands on your stick, you can strike him. You can also kick him.

In theory, one might reverse row under a grab, as though rowing backward. This is not as efficient as rowing forward, but it is better to know this option than not know it.

Solution Possibility 7: Weapon Hand-off
If the opponent grabs your weapon-bearing limb, you may be able to switch weapon hands.

Consider the basic common hand switch.

Solution Possibility 8: Charge! Weapon Recovery
Lose your weapon? Charge in immediately like a madman. Get your weapon back, etc.

Lose your weapon? Charge in like a madman and get it back before he can use it against you.

Your Review List for Weapon Retention:
  Retention 1: Proper pre-assessment of the situation and suspects.
  Retention 2: Proper retention holsters and sheaths.
  Retention 3: Core unarmed combatives (possibly the most important skills).
            -striking, kicking, blocking, grappling.

  Retention 4: The secure weapon is grabbed to establish a base for your next move.
  Retention 5: Release and step away. If your weapon limb is grabbed as you try to pull your weapon, let go of it, and take your hand away from the weapon.

  Retention 6: Basic releases
              - pull-aways and yank-frees.
              - circular releases clockwise or counter-clockwise.
              - slap and push-pull releases.
              - elbow roll-overs.
              - rowing.

  Retention 7: Weapon hand-offs to your free hand.
  Retention 8: Charge in! Weapon recovery. Lost it? Get it back.

Take the Weapon Retention Challenge!
Use the prior solutions and get your weapon back in force-on-force scenarios.

Single-handed grabs on your carry site
1: He is in front of you. He grabs your weapon on your carry site with his left hand.
2: He is in front of you. He grabs your weapon on your carry site with his right hand.
3: He is behind you. He grabs your weapon on your carry site with his left hand.
4: He is behind you. He grabs your weapon on your carry site with his right hand.
5: He is on your right side. He grabs your weapon on your carry site.
6: He is on your left side. He grabs your weapon on your carry site.
7: He is on top in a ground fight. He grabs your weapon on your carry site with his left hand.
8: He is on top in a ground fight. He grabs your weapon on your carry site with his right hand.
9: He is bottom-side in a ground fight. He grabs your weapon on your carry site with his left hand.
10: He is bottom-side in a ground fight. He grabs your weapon on your carry site with his right hand.

Single-handed grabs on your carry site, and he punches you
1: He is in front of you. He grabs your weapon on your carry site with his left hand. He punches.
2: He is in front of you. He grabs your weapon on your carry site with his right hand. He punches.
3: He is behind you. He grabs your weapon on your carry site with his left hand. He punches.
4: He is behind you. He grabs your weapon on your carry site with his right hand. He punches.
5: He is on your right side. He grabs your weapon on your carry site. He punches.
6: He is on your left side. He grabs your weapon on your carry site. He punches.
7: He is on top in a ground fight. He grabs your weapon on your carry site with his left hand, punches..
8: He is on top in a ground fight. He grabs your weapon on your carry site with his right hand, punches.
9: He is bottom-side in a ground fight. He grabs your weapon on your carry site with his left hand, punches.
10: He is bottom-side in a ground fight. He grabs your weapon on your carry site with his right hand, punches.

Double-handed grabs on your carry site
1: He is in front of you. He grabs your weapon on your carry site.
2: He is behind you. He grabs your weapon on your carry site.
3: He is on your right side. He grabs your weapon on your carry site.
4: He is on your left side. He grabs your weapon on your carry site.
5: He is on top in a ground fight. He grabs your weapon on your carry site.
6: He is bottom-side in a ground fight. He grabs your weapon on your carry site.

Single-handed grabs on your weapon limb as you draw
1: He is in front of you. He grabs your weapon limb with his left hand.
2: He is in front of you. He grabs your weapon limb with his right hand.
3: He is behind you. He grabs your weapon limb with his left hand.
4: He is behind you. He grabs your weapon limb with his right hand.
5: He is on your right side. He grabs your weapon limb on your carry site.
6: He is on your left side. He grabs your weapon limb on your carry site.
7: He is on top in a ground fight. He grabs your weapon limb with his left hand.
8: He is on top in a ground fight. He grabs your weapon limb with his right hand.
9: He is bottom-side in a ground fight. He grabs your weapon limb with his left hand.
10: He is bottom-side in a ground fight. He grabs your weapon limb with his right hand.

Single-handed grabs on your weapon limb as you draw. He punches.
1: He is in front of you. He grabs your weapon limb with his left hand. He punches.
2: He is in front of you. He grabs your weapon limb with his right hand. He punches.
3: He is behind you. He grabs your weapon limb with his left hand. He punches.
4: He is behind you. He grabs your weapon limb with his right hand. He punches.
5: He is on your right side. He grabs your weapon limb on your carry site. He punches.
6: He is on your left side. He grabs your weapon limb on your carry site. He punches.
7: He is on top in a ground fight. He grabs your weapon limb with his left hand. He punches.
8: He is on top in a ground fight. He grabs your weapon limb with his right hand. He punches.
9: He is bottom-side in a ground fight. He grabs your weapon limb with his left hand. He punches.
10: He is bottom-side in a ground fight. He grabs your weapon limb with his right hand. He punches.

Double-handed grabs on your weapon limb as you draw
1: He is in front of you. He grabs your weapon limb on your carry site.
2: He is behind you. He grabs your weapon limb on your carry site.
3: He is on your right side. He grabs your weapon limb on your carry site.
4: He is on your left side. He grabs your weapon limb on your carry site.
5: He is on top in a ground fight. He grabs your weapon limb on your carry site.
6: He is bottom-side in a ground fight. He grabs your weapon limb on your carry site.

Single-handed grabs on your single-hand gripped weapon
1: He is in front of you. He grabs your right-handed weapon limb with his left hand.
2: He is in front of you. He grabs your right-handed weapon with his right hand.
3: He is behind you. He grabs your right-handed weapon with his left hand.
4: He is behind you. He grabs your right-handed weapon with his right hand.
5: He is on your right side. He grabs your right-handed weapon with his left hand.
6: He is on your right side. He grabs your right-handed weapon with his right hand.
7: He is on your left side. He grabs your right-handed weapon with his left hand.
8: He is on your left side. He grabs your right-handed weapon with his right hand.
9: He is in front of you. He grabs your left-handed weapon limb with his left hand.
10: He is in front of you. He grabs your left-handed weapon with his right hand.
11: He is behind you. He grabs your left-handed weapon with his left hand.
12: He is behind you. He grabs your left-handed weapon with his right hand.
13: He is on your right side. He grabs your left-handed weapon with his left hand.
14: He is on your right side. He grabs your left-handed weapon with his right hand.
15: He is on your left side. He grabs your left-handed weapon with his left hand.
16: He is on your left side. He grabs your left-handed weapon with his right hand.
17: He is on top in a ground fight. He grabs your right-handed weapon with his left hand.
18: He is on top in a ground fight. He grabs your right-handed weapon with his right hand.
19: He is on top in a ground fight. He grabs your left-handed weapon with his left hand.
20: He is on top in a ground fight. He grabs your left-handed weapon with his right hand.
21: He is on bottom in a ground fight. He grabs your right-handed weapon with his left hand.
22: He is on bottom in a ground fight. He grabs your right-handed weapon with his right hand.
23: He is on bottom in a ground fight. He grabs your left-handed weapon with his left hand.
24: He is on bottom in a ground fight. He grabs your left-handed weapon with his right hand.

Single-handed grabs on your single-hand gripped weapon. He punches.
1: He is in front of you. He grabs your right-handed weapon limb with his left hand. He punches.
2: He is in front of you. He grabs your right-handed weapon with his right hand. He punches.
3: He is behind you. He grabs your right-handed weapon with his left hand. He punches.
4: He is behind you. He grabs your right-handed weapon with his right hand. He punches.
5: He is on your right side. He grabs your right-handed weapon with his left hand. He punches.
6: He is on your right side. He grabs your right-handed weapon with his right hand. He punches.
7: He is on your left side. He grabs your right-handed weapon with his left hand. He punches.
8: He is on your left side. He grabs your right-handed weapon with his right hand. He punches.
9: He is in front of you. He grabs your left-handed weapon limb with his left hand. He punches.
10: He is in front of you. He grabs your left-handed weapon with his right hand. He punches.
11: He is behind you. He grabs your left-handed weapon with his left hand. He punches.
12: He is behind you. He grabs your left-handed weapon with his right hand. He punches.
13: He is on your right side. He grabs your left-handed weapon with his left hand. He punches.
14: He is on your right side. He grabs your left-handed weapon with his right hand. He punches.
15: He is on your left side. He grabs your left-handed weapon with his left hand. He punches.
16: He is on your left side. He grabs your left-handed weapon with his right hand. He punches.
17: He is on top in a ground fight. He grabs your right-handed weapon with his left hand. He punches.
18: He is on top in a ground fight. He grabs your right-handed weapon with his right hand. He punches.
19: He is on top in a ground fight. He grabs your left-handed weapon with his left hand. He punches.
20: He is on top in a ground fight. He grabs your left-handed weapon with his right hand. He punches.
21: He is on bottom in a ground fight. He grabs your right-handed weapon with his left hand. He punches.
22: He is on bottom in a ground fight. He grabs your right-handed weapon with his right hand. A punch.
23: He is on bottom in a ground fight. He grabs your left-handed weapon with his left hand. He punches.
24: He is on bottom in a ground fight. He grabs your left-handed weapon with his right hand. He punches.

Double-hand grab on your single-handed gripped weapon
1: He is in front of you. He double-hand grabs on your right-handed grip weapon.
2: He is in front of you. He double-hand grabs on your left-handed grip weapon.
3: He is on your right. He double-hand grabs on your right-handed grip weapon.
4: He is on your right. He double-hand grabs on your left-handed grip weapon.
5: He is on your left. He double-hand grabs on your right-handed grip weapon.
6: He is on your left. He double-hand grabs on your left-handed grip weapon.
7: He is on the top of a ground fight. He double-hand grabs your right-handed grip weapon.
8: He is on the top of a ground fight. He double-hand grabs your left-handed grip weapon.
9: He is on the bottom of a ground fight. He double-hand grabs your right-handed grip weapon.
10: He is on the bottom of a ground fight. He double-hand grabs your left-handed grip weapon.

He takes. You recover. (Unarmed Combatives!)
1: He takes your stick from the front. You recover it.
2: He takes your stick from the right side. You recover it.
3: He takes your stick from the left side. You recover it.
4: He takes your stick from behind. You recover it.

Chapter 15: The Stick Duel
Stick versus Stick? Really?

It is unlikely that citizens, police, or the military will be involved in any prolonged stick versus stick fighting in many countries. This is mostly a martial arts event and is not the main purview of a tactical stick program. You can still learn some abstract skills by doing this, develop footwork and, in most cases, a much greater depth in stick handling. Many of the newer police tactical baton programs now ask their attendees to experience soft-stick sparring with pads for this reason.

In fact, even the world's most pervasive stick-fighting arts, the Filipino Martial Arts (FMA) that emphasize stick versus stick fighting - are actually doing abstract training. The FMA stick is representative of the machete, or bolo. So even when FMA practitioners are stick fighting, they are really symbolically edged-weapon fighting. Still, through the years, stick fighting itself has become elevated into is own entity, and many ignore the edged-weapon roots and approach the techniques as sticks.

In the mixed-weapon world we live and work in, most likely you will be fighting against a weapon unlike yours in some sort of duel, if a so-called duel will actually develop and not an ambush. All fights are highly situational. This could be a baseball bat versus a crowbar, or a golf club versus a tire iron. A tree limb versus a shovel. A pool cue versus a pool cue. Or yes, maybe even a stick versus a stick. Fighters may phase in and out of duels. So, a percentage of proper, comprehensive training still must include impact weapon versus impact weapon training and doctrine.

Run the probability numbers in your head, and plan for the "who, what, where, when, how, and why" possibilities. Though the odds are slim, a properly trained impact weapon practitioner also trains for this very dueling event in a proper proportion to reality.

Even when opponents fight stick versus stick in today's world, it is not usually a duel like a sword or fencing match as one might see in the classic movies. The encounters are highly situational. In the photos above, it's stick versus stick...and pepper spray gas, too!

The great Filipino Arnisador Remy Presas, veteran of many competition stick fights and real world fight challenges once said, *"you train your whole life for a four second stick fight."* It doesn't take much to hit a head or a key target and end the fight. Stick dueling can be like sword fencing and sword dueling, and then it is not. An impact weapon is a different weapon.

If you prepare for the fast-action of a impact weapon duel, you need both the skills mentioned in each section of this book and ones introduced here in this chapter.

In summary, as a dueler, you will need:

1: Footwork and body evasion skills - to move in and out, side-to-side, up and down.

2: Striking skills - to strike fast, powerfully and efficiently. This also includes support strikes and kicks.

3: Blocking skills - at times in these rapid fire encounters, rapid blocks look like and feel like strikes, yet it is good also to learn solid blocking skills. Blocks will happen.

4: Counters to Common Blocks - the four Ps. Pinning, passing, pulling or pushing.

5: Fakes, tricks and set-ups - to evoke, lure and trick the enemy.

6: Endurance - to last in such a high tension extended situation.

7: Tenacity and pain tolerance - to last through the encounter.

8: Proper safety gear. You don't need to be shot by a bullet to know the bullet hurts you. Nor do you need to be beaten by a real stick, your body unprotected, to learn how to train for a stick duel. You need helmets, gloves and body padding. You may even use softer sticks. The safer your training, the longer you train, the more you learn. Invent ways to stay safe.

9: The great "etc."...- all these fights are situational. Be prepared for the chaos.

If you are in a stick-versus-stick duel, one of your main priorities is to pick something else up and fight with it. Use something as a shield or a second impact weapon

from your environment. Absent all other options, and you are in the theoretic duel. Here is how you hold an opponent at bay, or set him up for attack. Just as a boxer or kick boxer may conceal their attacks with moving hands, the stick fighter does the same with a moving stick.

Your Stop 1 stalking, ready-moves could contain all the stick strikes and support strikes and kicks we've covered here in this book.

Many stick fighters like to stand still, stick raised and suddenly swing. Fakes with the stick and with footwork, such as sudden quick steps, even yelling bursts, are also favorites.

Some use all of the above in a single sparring session. Some use half or just a few tricks while dueling. Some will suddenly yell out and stomp their feet.

After some practice and coaching, you will develop a style, like a boxer does, that best suits your body shape and speed, within the law and/or the rules of engagement (if any.).

Always, always, always get something else to fight with as a shield or another impact weapon. Always get the advantage.

Stop 1 Dueling Footwork and moves

Single slashes
Multiple stashes side-to-side
Multiple slashes up and down
Multiple X cuts
Big circles at the elbow and shoulder
Small circles at the wrist both inside and outside the arm
Shaft and tip extensions that pump:
> * high.
> * medium.
> * low.
> * high, low.
> * low, high.
> * high medium.
> * random combinations thereof.

Stand still, stick raised, and swing.
Footwork.
Fakes with the stick and with suggestive footwork.
Combinations of some or all of the above.

Here is a low, shaft pump at the opponent.

The Stick Fake

Used in stick dueling and is a universal strategy in striking with an impact weapon, we must include this move in non dueling also. In just about every competitive sports to fighting to world wars there are fakes and deceptions. In the realm of impact weapons, aspects of the "stick fake" is an important tactic whether you are a cop on the street corner with a night stick, or a Filipino stick sparrer. In the realm of impact weapons the "stick fake" is an important strategy. Oddly, there are some people who say never fake, or that fakes don't work. I don't believe this.

There are micro (small) to macro (big) fakes in all sorts of disciplines. They create openings for attacks. Some systems will make faking very complicated but, if you embrace the Combat Clock, you can teach the skills of faking very simply, quickly, and in an organized fashion. Here is an easily understood example. You look as though you are going to strike high and from above. You rear back your weapon bearing limb. You get the seduced, elicited response you wanted. He raises his arms and weapons to block your high attack.

You look like you are striking down from high or 12 o'clock....

You next swing your arm around in somewhat of a circle backward and hit him in the knees or groin. I have done this particular fake numerous times while stick fighting as well as in law enforcement, and it has always worked.

This would be called a 12 o'clock fake to a 6 o'clock strike in our Combat Clock vernacular. Now imagine the combinations. Create your own workout list and develop these moves. Remember your fake depends upon the response time of your opponent. If he is very slow, then your fast fake is too fast, and he hasn't the time to be seduced into position. Pace the fake per your opponent.

...swing your arm around back and strike the knee or groin.

Basic Training Fake Workout

Fake at 12 -   strike at 12 - after his block passes
                strike at 3
                strike at 6
                strike at 9
Fake at 3 -   strike at 3 - after his block passes
                strike at 6
                strike at 9
                strike at 12
Fake at 6 -   strike at 6 - after his block passes
                strike at 9
                strike at 12
                strike at 3
Fake at 9 -   strike at 9 - after his block passes
                strike at 12
                strike at 3
                strike at 6

Advanced Training Fake Workout
Use all 12 numbers of the clock in the basic manner.

I do not wish to over-emphasize stick versus stick dueling, but I will at leats mention these tips as the event may occur with baseball bats, axe handles, tire irons, etc.

Duel Point 1: You will be in athletic, sports motion.
Duel Point 2: Stick may be held still and strikes suddenly.
Duel Point 3: Stick may travel in a figure eight motion.
Duel Point 4: Stick may go up and down.
Duel Point 4: Stick may go in and out.
Duel Point 5: Stick may go side-to-side.
Duel Point 6: Dueler uses one, or some or all of the above.
Duel Point 7: Faking as expressed through the Basic 4 or Advanced 12 Combat Clock.
Duel Point 8: Lures. The stick retracts in a poor defensive position to set up a predictable counter-attack, such as:
- Lure 1: Open your belly, hit his incoming attack.
- Lure 2: Offer stick hand, hit his incoming attack.
- Lure 3: Offer free hand, hit his incoming attack.

Duel Point 9: The support hand works with these motions:
a) holds another weapon as an:
- knife?
- shielding device?
- projectile?
- may work opposite the stick movements.
- may work with the stick movements.

b in a guarded position independent of the movements.
c) flicks and fans for distraction.

Duel Point 10: The Diminished Fighter Theory (DFT) The opponent may tire from wounds or lack of fitness.

My first serious Filipino instructor Ray Medina and me stick sparring-dueling.

Two Hand Grip Dueling - The Military Pugil Course

This set of instruction, long a staple of training for many militaries, is vanishing from military doctrine. I have enclosed it in this section from my old Army manuals for practical information as well as historical significance. The added photos are taken from my personal collection from my time in the army. This is very valuable, hard core drill training. This is how it was done, and how you might do it again.

PUGIL TRAINING

Training in pugil techniques prepares the soldier to confidently and aggressively use the rifle-bayonet. It furnishes the rifle-bayonet fighter with an opponent who can think, move, evade, fight back, and (most important) make corrections. It provides realism.

Section I
EQUIPMENT

Pugil equipment consists of the pugil stick and protective gear that is especially designed to protect the soldier during training. It allows the soldier to participate in pugil training without incurring or fearing injury. Participation with no fear of injury helps the soldier to develop an individual style of fighting and improve his ability to fight with the rifle and bayonet. Pugil equipment (Figure A-1) is designed to prevent injuries to the head and face, chest, groin, and hands.

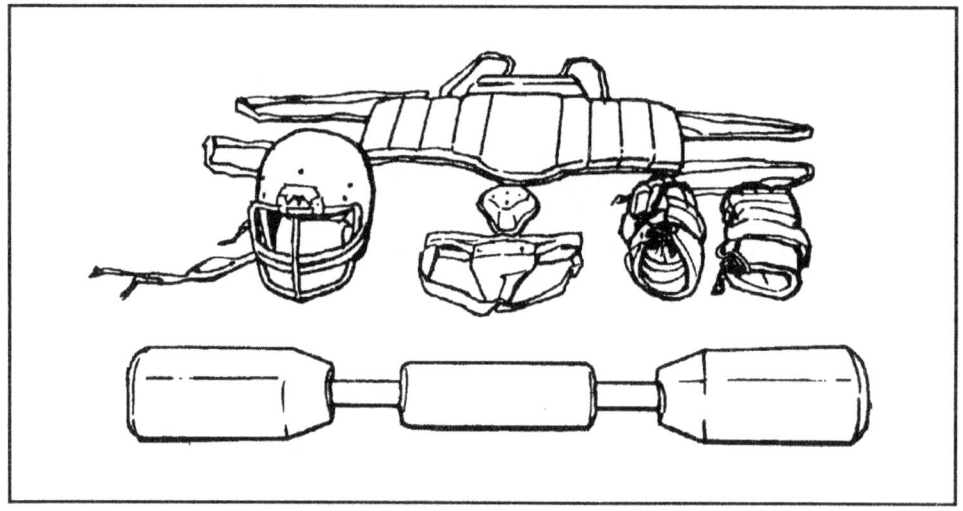

Figure A-1. Pugil equipment.

A-1. SECURING EQUIPMENT

Units can construct pugil sticks or obtain them from the Training Support Center. The helmets with attached face masks, gloves, chest protectors, and boxers' protective cups are nonstock-type commercial items. Locally used nonstandard stock numbers identify these commercial items, which are obtained through TSC or local purchase.

A-2. HEADGEAR

Headgear consists of a regulation football helmet with a face mask attached (Figure A-2). When purchasing these helmets, you should consider the varying head sizes of individuals. For each 100 helmets purchased, it is recommended that 10 percent be 6 1/2 to 6 3/4 in size, 80 percent be 6 7/8 to 7 1/8 in size, and 10 percent be 7 1/4 to 7 1/2 in size. Adjust helmets that are too large for an individual by adding foam rubber to the inside of the helmet. To secure the helmet to the head, use a chin strap made of vinyl plastic and foam rubber.

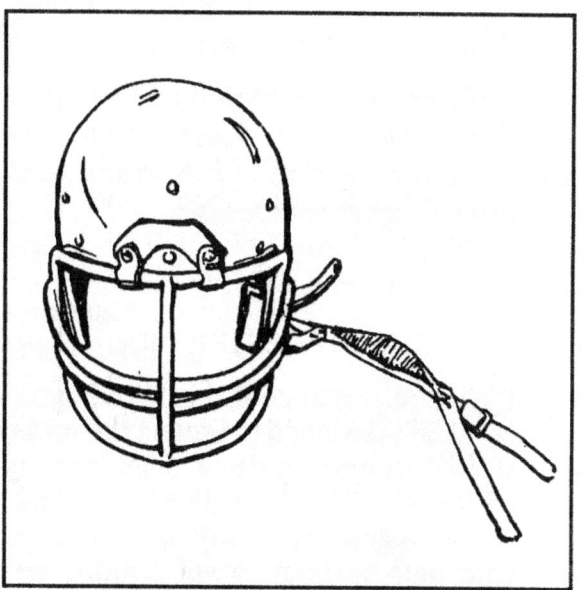

Figure A-2. Pugil headgear.

A-3. GROIN EQUIPMENT

A boxer's protective cup of the variety used in athletic competition protects the groin (Figure A-3).

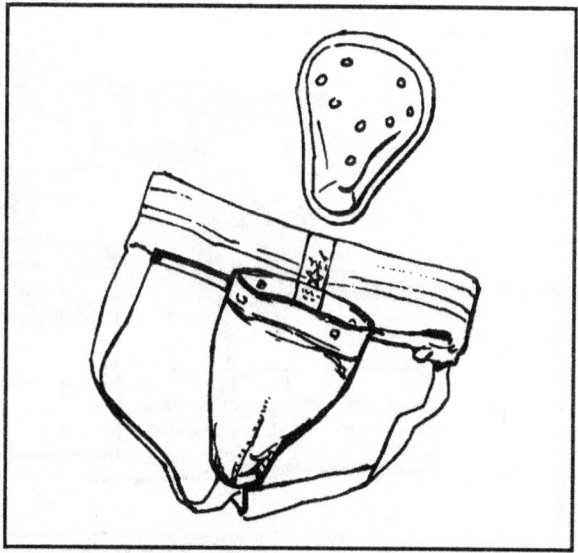

Figure A-3. Boxer's protective cup.

> **CAUTION**
>
> GLOVES AND CHEST PROTECTORS MUST BE WORN DURING TRAINING.

A-4. HAND EQUIPMENT

Gloves are required in pugil training. Hockey gloves (Figure A-4) provide maximum protection for the fingers and joints of the hands and wrist and aid in controlling the stick.

A-5. CHEST PROTECTORS

Soldiers must use chest protectors (Figure A-5) during pugil training to prevent injuries. Baseball catchers' chest protectors or martial arts protectors are recommended. If chest protectors are unavailable, substitute flak vests.

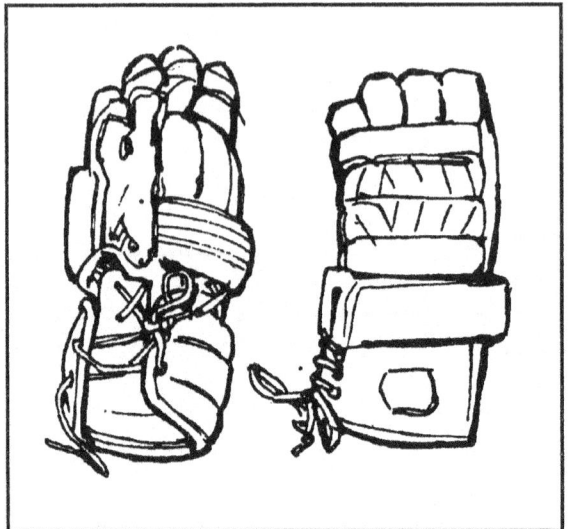

Figure A-4. Hockey gloves.

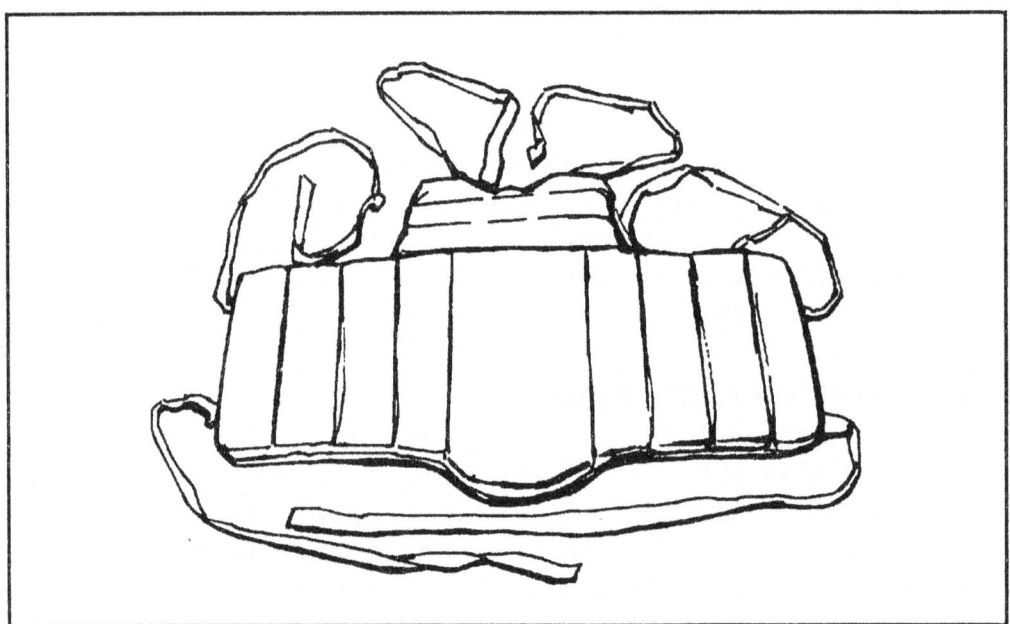

Figure A-5. Chest protector.

A-6. PUGIL

Pugil sticks (Figure A-6) may be obtained from the local TSC or call Devices Section, TSC, Ft Benning, GA, DSN 835-1407.

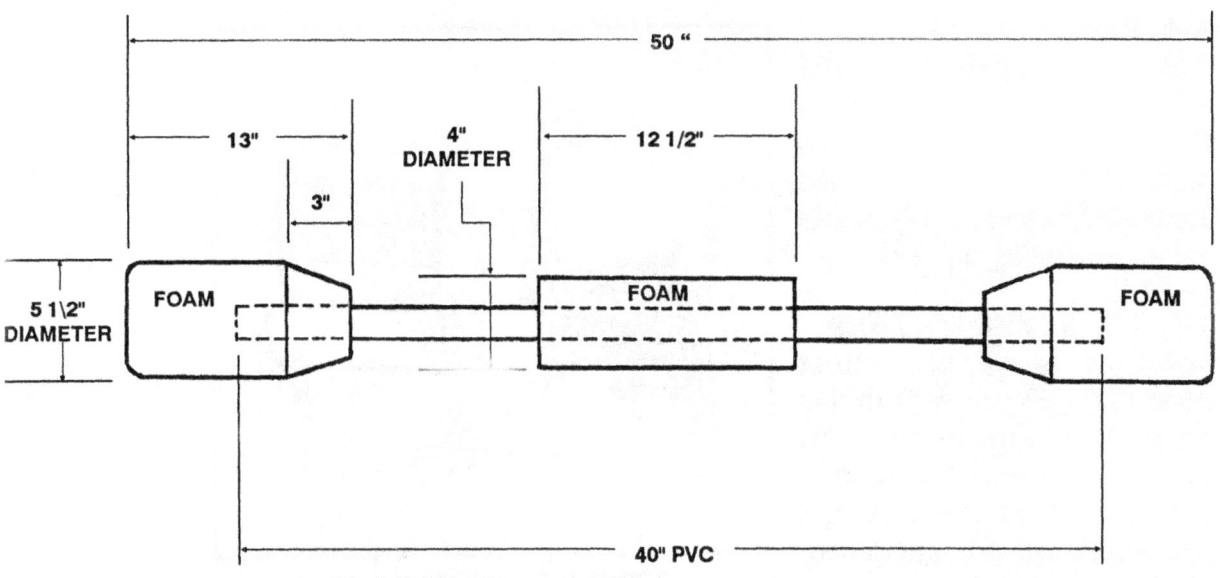

Figure A-6. Materials in pugil stick construction.

Section II
TECHNIQUES

Pugil training is a way to teach the soldier to use the rifle-bayonet with confidence and aggression. After the soldier becomes skilled in the basic positions and movements with the rifle-bayonet, he should be introduced to pugil training techniques.

A-7. VALUE OF PUGIL TRAINING

Since pugil fighting is a rugged contact activity, the soldiers must remain alert. They act and react from instinct, thus affording an opportunity to develop their individual rifle-bayonet fighting skills. Little effort is required by the instructor to motivate the soldiers—the pugil stick is the motivating force. Soldiers derive much physical benefit from pugil training, and they develop an aggressive mental spirit that is so essential if the rifle-bayonet fighter is to be successful in combat.

Figure A-7. Man-to-man contact.

    a. The instructor must consider several factors to gain maximum effectiveness from pugil training. These include training, control, supervision, safety, and protective equipment.

    b. The instructor should teach the rifle-bayonet fighter the basic positions and movements, as well as the series of follow-up movements, with the rifle-bayonet before beginning pugil training.

c. The pugil stick should approximate the length and weight of the M16 rifle with bayonet attached for maximum training benefit. Substitution of the pugil stick for the rifle provides an opportunity to improve skill and test each soldier's ability to perform against a realistic, evasive target. All the positions and movements with the pugil stick are the same as with the rifle and bayonet (Figure A-8).

A-8. CONTROL, SUPERVISION, AND SAFETY
Instructors supervising pugil training must understand its values and limitations. The instructor maintains control of the bout at all times; his best method of control is by blowing a whistle to start and stop action. He is alert to prevent wild swinging of the pugil sticks, and he ensures that the soldiers keep their eyes on each other. For safety reasons, he should pair soldiers who are about the same height and weight.

Figure A-8. Substitution of the pugil stick for the rifle.

a. Soldiers use only the positions and movements that they have been taught in rifle-bayonet training. They must hold the stick and deliver blows as if using the rifle-bayonet.

b. One instructor is necessary for each bout; however, he needs assistance to supervise the fitting and exchanging of equipment. The instructor makes sure the equipment fits properly and watches constantly for any loose or broken equipment. As soon as he sees any insecure equipment, he stops the bout to prevent possible injuries. After deficiencies have been corrected, the round is resumed.

c. The instructor insists that the soldiers growl during the bouts; this adds to their aggressiveness and tends to reduce tension.

d. Soldiers with medical problems, such as hernias, frequent headaches, previous brain concussions, recent tooth extractions, or lacerations with stitches, must be excluded from pugil training for safety reasons. Therefore, before conducting pugil training, it is necessary to determine if anyone should be eliminated from participation. Finally, instructors should always be alert for the unexpected and, if in doubt, stop the bout immediately to prevent injury.

A-9. WARM-UP ROUND

In the early stages of pugil training, maximum benefit is gained by working with platoon-size groups (or smaller) in a circular formation. Two soldiers engage in a pugil bout in the center of the circle. An instructor critiques them so all soldiers can learn from observed mistakes. The soldiers assume the attack position 12 steps from each other. In the first round, the instructor allows them freedom of movement to prove to soldiers that the equipment provides ample protection from a hard blow. Everyone should take part in as many bouts as necessary to gain skill before going on to more advanced training. Immediately after the warm-up round, the soldiers engage in graded bouts.

A-10. GRADED BOUTS

During graded bouts (Figure A-9, page A-8), the opponents face each other, 12 steps apart. The instructor should be in a position where he can best control the bout. Each bout consists of three rounds. To score a point or win a round, a soldier must score a solid blow with either end of the pugil stick to a vulnerable point—the head, throat, chest, stomach, or groin region.

a. To start a bout or a round, the instructor blows the whistle, and the soldiers move toward each other in the attack. The instructor awards one point to the soldier striking the first disabling blow. A disabling blow is any blow that is delivered to a vulnerable part of the opponent's body. When a soldier strikes such a blow, the instructor uses a whistle to stop the round. At the end of the round, soldiers move back to their respective lines, assume the attack position, and wait for the signal to start the next round. The soldier who wins two out of three rounds wins the bout.

b. The instructor should encourage soldiers to move in aggressively and to attack violently, using any of the attack movements learned during rifle-bayonet training. If the soldier misses or his opponent sidesteps, he should immediately follow up until he has landed a blow to a vulnerable spot.

c. The soldier who hesitates to strike his opponent realizes that defeat can be quick; therefore, he tries to be aggressive and overcome his opponent in the shortest possible time.

d. Because training is done in two-man bouts, a squad, platoon, and finally a company champion may be selected. The instructor should encourage competition throughout the pugil training program.

A-11. PUGIL COURSES

After several two-man bouts, the rifle-bayonet fighter is ready for the human thrusting target course and the human thrusting assault course.

a. **Human Thrusting Target Course.** Eight to ten soldiers are lined up in file formation, 12 steps apart. The instructor selects each soldier to act as a specific-type target. The rifle-bayonet fighter, also in pugil gear, walks to each human target, moving with the pugil stick at the attack position. As the rifle-bayonet fighter approaches an opponent, the opponent shouts the movement that the rifle-bayonet fighter is to execute—for example, thrust, slash, butt stroke. After executing the movement, the rifle-bayonet fighter pauses long enough for the instructor to make corrections, then he moves to the next target. The number of walk-throughs depends on each soldier's ability to execute the movements correctly. Next, he runs through the course at full speed, growling and executing the called movements with maximum force against his opponents. The duties are rotated so that all soldiers get to act as fighters and as human targets.

    b. **Human Thrusting Assault Course.** A qualification-type course can be conducted to measure each soldier's skill. This course should approximate an obstacle course in length, obstacles, and terrain. The course layout should take advantage of natural obstacles, such as streams, ditches, hills, and thickly wooded areas. Soldiers in pugil equipment can be placed among the obstacles to act as human targets. The rest of the unit, in pugil equipment, can negotiate all obstacles and human targets, using instinctive rifle-bayonet fighting movements.

Chapter 16: Impact Weapon Takedowns

Knock 'em down,
Push 'em down,
Pull 'em down,
Trip 'em down,
Twist and turn 'em down

Grappling is an important combat skill. A stick in one hand changes the unarmed combatives paradigm considerably. Empty-handed, a citizen, enforcement officer or soldier takes the enemy down to the ground by knocking him down, pushing him down and/or pulling him down, twisting and turning his head, body or limbs, or tripping his legs. Most times, two free hands are an advantage in grappling and takedowns, and a necessity while executing more than half the official list of universal takedowns. But taking the enemy down while holding a stick in a saber or reverse grip in one hand changes the classic, grappling battle plan. The soldier loses one gripping hand. He must hook the body part, which is not as solid as a grip. He must push, pull and turn with the stick. He must learn not to use a percentage of common, two-handed takedowns and throws.

Given training time and training facility constraints, it becomes impossible to train large police or military units in long-term, high-level, grappling skills in an effort to create a virtual black belt skill level in "jujitsu" takedowns. The following training outline is a simple, scientific breakdown of basic grappling takedowns while holding a stick.

In order to execute takedowns while holding a stick, the practitioner needs to practice body rams, empty hand striking skills, stick striking skills, support and weapon-bearing limb striking skills, grabbing and wrapping skills, and off-balancing and tripping skills. Rather than face the challenge of describing every takedown while holding a stick, which would be the subject matter of an entire thick book by itself, the following chapter is an overall synopsis. This synopsis added with the upcoming combat scenario chapters will provide a soldier with a healthy working knowledge of stick takedowns.

Stick Combat Takedown Overview

Stick Takedown Study 1: "Knock 'em down!"
Knocking the enemy down means using all support hand strikes and all kicks to knock the enemy off of his feet.

Stick Takedown Study 2 and 3: "Push 'em down!" and/or "Pull 'em down!"
Pushing, pulling, or pushing and pulling the enemy down means grabbing and hooking body parts and exerting shoves, pulls or shoving and pulling to trip an enemy off his feet.

Stick Takedown Study 4: "Twist and turn 'em down!"
Twisting the enemy down means twisting the enemy's wrists, arms, head, torso or legs until he loses balance, trips and falls down.

The Takedown Geography Map

The takedown collection will include a study of stick applications upon body parts - the head, neck, torso, arms and legs. On the arms we include the upper arms, elbow, lower arm, wrist and hand. On the legs we look at the pelvis, upper leg, knee, lower leg, ankle and feet. Each of the aforementioned has multiple takedown potentials when at the mercy of a striking, pulling, pushing or turning stick.

Method 1: The strike takedown
If you strike the head, neck, clavicle, wrist hand and the knee area it may drop an enemy. Many enforcement officials are taught to strike on the nerve of the outer thighs as a major takedown target, but many stick versus stick fighters receive strikes there all the time and do not fall. But, any significant stick strike anywhere else may have the potential of shutting down an opponent.

Method 2: The push takedown
This method pushes the enemy over. Sometimes you need to trip/trap a leg or legs, and sometimes not, as you might find success with a push anywhere.

Method 3: The pull takedown
This method pulls the enemy over. Sometimes you need to trip/trap a leg or legs, sometimes not, as you might find success with a pull anywhere.

Method 4: The turn/twist takedown
This method turns and twists the enemy over. Sometimes you need to trip/trap a leg or legs, sometimes not, as you might find success with a turn anywhere.

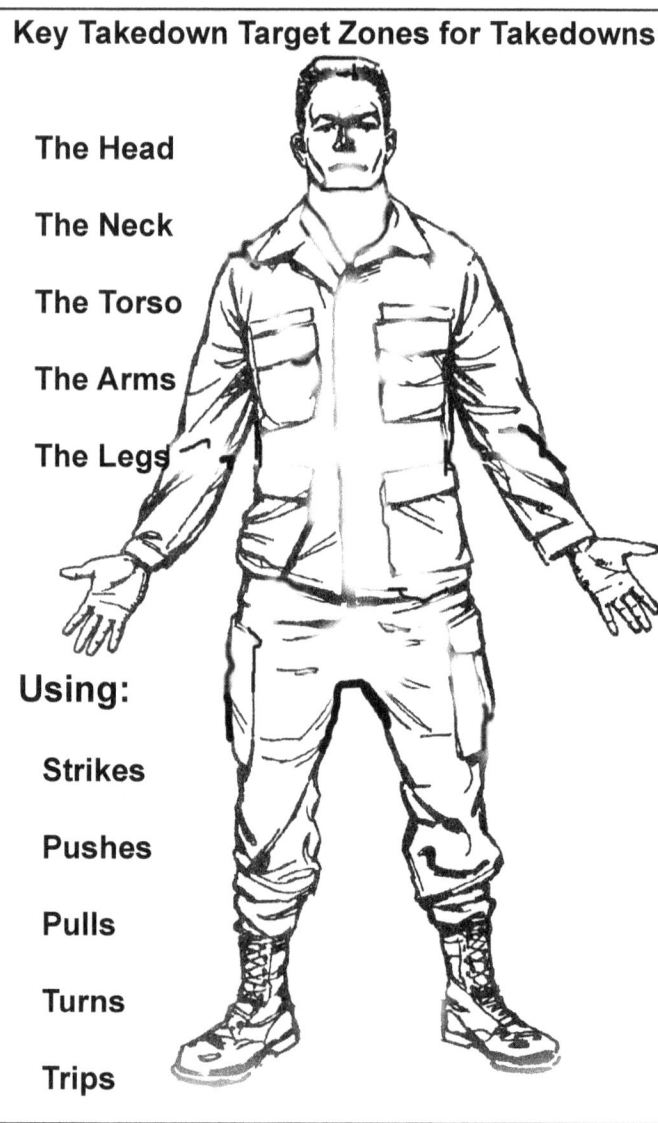

Key Takedown Target Zones for Takedowns

- The Head
- The Neck
- The Torso
- The Arms
- The Legs

Using:
- Strikes
- Pushes
- Pulls
- Turns
- Trips

In the chaos of a fight, you will learn that it often takes a mix of the above movements to get someone down. This chapter will also include classic enforcement escorts organized to fit into the anatomy categories. It will also include an overview on ground fighting with a stick in your hand.

"Tug of Wars" Warm-ups

These are not completely tugs of war, meant to pull your partner over a particular finish line. Rather, they are studies in energy and methods to feel and outsmart the push and pull of your partner. These appear in no special order. The drills are great warm-ups to start any impact weapon class. Also, they should all be done with replica long guns. Either way, these force-on-force sensitivity exercises set up many grappling situations.

Drill 1: The Two Hand Grip Riot Push/Pull

Both parties grab one stick held in the riot position. Both pull and push in a tug of war balance and positioning battle. You are not simply trying to pull your partner over an imaginary line. You also push when he pulls, pull when he pushes. The end goal is to experience the push/pull of fighting over the weapon.

Drill 2: Double Stick Push/Pull

Both parties grab two sticks held in each hand as shown to the left. Both pull and push in a tug of war balance and positioning battle. You are not simply trying to pull your partner over an imaginary line. You also push when he pulls, pull when he pushes. The end goal is to experience the push/pull of fighting over the weapon.

Drill 3: Two Hand Grip Interlock Push/Pull

Both parties grab sticks. Both interlock their arms. Both pull and push in a tug of war balance and positioning battle. You are not simply trying to pull your partner over an imaginary line. You also push when he pulls, pull when he pushes. The end goal is to experience the push/pull of fighting over the weapon. Take care in this one not to injure your partner's torso. You will learn here just how easily this can be done.

Drill 4: Stick vs. Unarmed Combatant Push

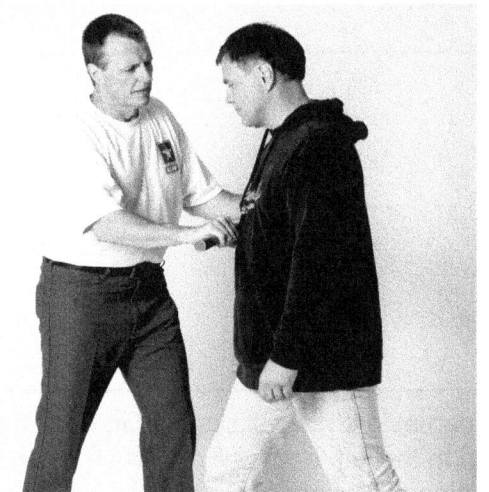

This is a study to see how powerful a two-hand shove can be. Push low? He will usually bend over first, then move back. Push high? He will bend first, and then move back. Push in the zone between the elbow and the shoulder, and you will have an instant shove. As with all these drills, use your full body synergy to get the job done. Push the front, the back, the right, the left.

Drill 5: Stick versus Stick Port Arms Push

This is push versus the port arms position. Check the flexibility of your chosen weapon. See how pushing works out.

Using his stick (or arm) as an axis point turn an end in to see what you can strike, hook, pass, switch to one hand and catch a limb? Experiment.

An opponent should be sufficiently stunned before a takedown attempt.

But First, the Combat Scenario Set-ups for Takedowns...

The Statue Drill study.

How should we or do we get to the point of a takedown? In order to get into a position for a takedown, you must have some preliminary action. Your opponent could be empty-handed, holding an improvised weapon, holding an impact weapon, or a knife or be drawing or presenting any firearm. Some diverse, common scenario modules for the set-ups are the *Spartan Drill*, the *Chain of the Stick Drill* and the *In The Clutches of Drill.*

All of these prepare you for defense, offense, stunning and finally positioning. For positioning, you will use foot work and maneuvering to be outside his left arm or his right arm, inside his arms, or behind him. To illustrate a great way to teach new students these ideas is to introduce them to the classic martial arts statue drill.

 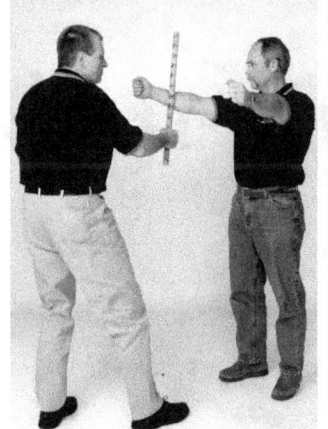 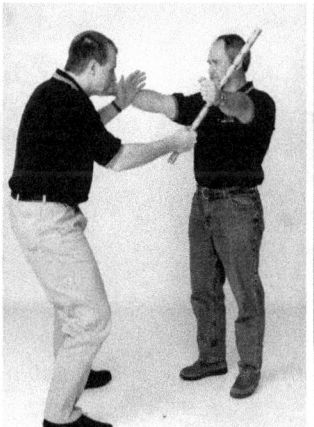 

Review the outside, inside, split, inside, outside movements.

3 Combat Scenario Set-up Drills
1: *The Spartan Drill* - The impact anywhere causes a disarm.
2: *The Chain of the Stick Drill* - The impact fails to cause a disarm. You grab his weapon bearing limb.
3: *In the Clutches Of Drill* - Both fighters grab each other's limb.

Positioning for a Takedown
You will be moving to be:

- Outside his right side
- Inside his right arm
- Inside his left arm
- Outside his left arm
- and behind him

The Spartan Combat Drill

The three combat drills work together in a succession of possibilities, and the first one is the Spartan Drill. This drill is all about the successful impact disarm. The other two drills are when the impact fails to disarm. In the Spartan Drill, you strike the enemy's weapon bearing limb. This could be someone attacking you with hand strikes and kicks, or a stick, a knife or drawing and/or presenting a forearm. Once the weapon has been successfully disarmed from this impact, you stun the enemy and take him down. The equation is block/strike the limb, stun, takedown. The takedowns appear later.

The impact could be anywhere and create a body-shocking disarm. I recall the story of one man who was hit on the empty-handed, left arm so hard that a shocking pain shot across his upper back, rippled through there and injured his shoulder muscles and caused him to drop the stick in his right hand. Of course, a blow to the head can disrupt the brain and cause a disarm.

The attacker strikes.   *You move and block/strike the limb.*

The drill also covers blasting the limb attempting to pull a weapon from a weapon carry site either on or off the body. Or, bashing someone holding a weapon on you or on someone else.

The takedowns themselves appear in the upcoming pages.

The impact disarm worked!   *You stun the enemy to take him down.*

Your impact weapon interrupts weapon quick draws and weapon presentations.

The Spartan Drill Progression Exercise Series:

Interrupting Quick Draws
   From the primary, secondary, and tertiary weapon carry sites, the trainer attempts to draw:
   - a knife, and the trainee strikes the limb.
   - a stick, and the trainee strikes the limb.
   - a handgun, and the trainee strikes the limb.
   - related expedient weapons, and the trainer strikes the limb.

Interrupting Weapon Threats and Presentations
   Strike the weapon limb of a knife, stick, firearm threat presentation (weapon out).
   - upon yourself.
   - upon others.

While being attacked
   The trainer attacks on the Basic Training 4 Combat Clock corners.
   - 12 o'clock slash.   - 12 o'clock stab.
   - 3 o'clock slash.   - 3 o'clock stab.
   - 6 o'clock slash.   - 6 o'clock stab.
   - 9 o'clock slash.   - 9 o'clock stab.

   The trainee
   Step 1: Dodge/block/strike the incoming limb and the weapon drops.
   Step 2: Strike to stun (or knock down) several times as needed.
   Step 3: Execute an efficient and proper takedown fitting the position you and he are in (takedowns appear later in this chapter).

Exercise through the Advanced 12 numbers of the Combat Clock with the above.

Chain of the Stick Combat Drill

The three combat drills work together in a succession of possibilities, and the first one is the Spartan Drill. This second drill is Chain of the Stick and is all about an unsuccessful impact disarm. As in the Spartan Drill, you strike the enemy's weapon bearing limb. This could be someone attacking you with a stick, a knife or drawing and/or presenting a forearm. This time the weapon limb is struck, but the weapon is not dislodged. The chain begins as you must grab the weapon-bearing limb. The equation is dodge/block/strike the weapon limb, grab the weapon limb, stun, takedown. The grab is the first linking grab of survival, ergo the chain name. Takedowns appear later in this chapter.

The attacker strikes.

You move and block/strike the limb.

The impact disarm did not work! You must ...

...grab the limb and strike quickly.

The Chain equation is block, grab, strike

Block - stopping the attack. but no impact disarm!

Grab - securing the blocked attack
* grab limb.
* grab weapon.

Strike - to stun
Follow-up - whatever is needed, such as a takedown.

The Chain Drill Progression Exercise Series:

Interrupting Quick Draws
  From the primary, secondary, and tertiary weapon carry sites, the trainer attempts to draw:
  - a knife, and the trainee strikes the limb, the disarm fails and he grabs the limb.
  - a stick, and the trainee strikes the limb, the disarm fails and he grabs the limb.
  - a handgun, and the trainee strikes the limb, the disarm fails. He grabs the limb.
  - related expedient weapons and the trainer strikes the limb. The disarm fails and he grabs the limb.

Interrupting Weapon Threats and Presentations
  Strike the weapon limb of a knife, stick, firearm threat presentation (weapon out)
  - upon yourself.
  - upon others.

While being attacked
  The trainer attacks on the Basic Training 4 Combat Clock Corners
  - 12 o'clock slash
  - 3 o'clock slash
  - 6 o'clock slash
  - 9 o'clock slash
  - 12 o'clock stab
  - 3 o'clock stab
  - 6 o'clock stab
  - 9 o'clock stab

  You the trainee
  Step 1: Dodge/block/strike the incoming limb, and the weapon does not drop.
  Step 2: Your support hand grabs the weapon bearing limb.
  Step 3: You strike the head or other vital target to stun.
  Step 4: Execute an efficient and proper takedown fitting the position you and he are in (takedowns appear later in this chapter).

  Exercise this 4 step Chain Format
  - versus empty hand strikes and kicks.
  - versus knife attacks.
  - versus stick attacks.
  - related expedient weapons.
  - add in a need to strike/block at 1/2 beat, empty hand strike.

  Exercise through the Advanced 12 numbers of the Combat Clock with the above.

The In the Clutches Of Combat Drill

The three combat drills work together in a succession of possibilities, and the first one is the Spartan Drill. This second drill is Chain of the Stick and is all about an unsuccessful impact disarm. As in the Spartan and Chain Drills, you strike the enemy's weapon bearing limb. (This could be a stick or a knife?) Once the weapon limb is struck, the weapon is not dislodged as in the Chain Drill. You grab his weapon-bearing limb. As you try to strike to stun, in this set he grabs your weapon, or your weapon-bearing limb, and you are in the clutches of each other. These clutches hand grab configurations are hand-on-forearm, hand-on-wrist, hand-on-hand and hand-on-stick and hand-on-neck.

Hand grab on forearm.

Hand grab on wrist.

Hand grab on hand.

Hand grab on stick

Grabs.

For the purpose of our study, the "In the Clutches Of" grabs are:
- hand on forearm.
- hand on wrist.
- hand on hand.
- hand on stick or gun.
- hand on neck, etc.

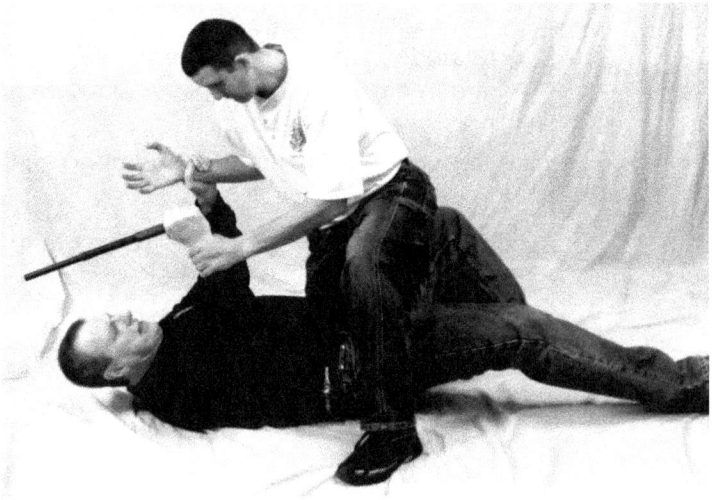
He has you. You have him In the Clutches Of each other.

The Clutches exercise samples are:

Clutches 1: High in the clutches of - both your gripping hands are high.
Clutches 2: Low in the clutches of - both your gripping hands are low.
Clutches 3: Split in the clutches of - one grip is high, the other low.
Clutches 4: Right side - both gripping hands are more on the trainee's right side.
Clutches 5: Left side - both gripping hands are more on the trainee's left side.
Clutches 6: Accordion clutches - both are virtually chest-to-chest as the arms. and grippings hands are spread out. At times, even chest to chest.

Clutches 7: Righty versus lefty - the arms are crossed, one grip high, one low.
Clutches 8: Some of the above apply to the ground.

Escape his clutches with the very same skills extensively covered in the weapon retention chapter. Use circular releases, elbow roll-overs, push-pulls, kicks to the groin and other methods used in that retention chapter.

High In the clutches Of.

Low In the clutches Of.

Split In the clutches Of.

Right side In the clutches Of.

Left side In the clutches Of.

Accordion clutches.

Righty vs. lefty.

Ground versions.

Basic releases

- pull-aways and yank-frees.
- circular releases clockwise or counter-clockwise.
- slap and push-pull releases.
- elbow roll-overs.
- rowing.
- One unique release not previously shown is one "righty versus lefty" solution - *The Arm Ram Release.* Ram your arm against his to gain a release.

Been grabbed? Use It!
When your weapon limb wrist or forearm has been grabbed, use the impact weapon as a strike.

Do you have freedom of movement in your wrist?

Fan strike the head.

Hit the arm as you do a circular release.

The Trap and Release Sample Solution
This is a combat scenario worth practicing to engrain the movements. Similar situations may arise. It involves arm-wrap-trapping his arm and, when your weapon hand is grabbed, you may bring the capture near your free hand. Grab him. Try a push/pull release and capture of that limb. You might even bait the capture of your hand to ensnare it and allow for unobstructed strikes to the face.

You wrap his arm. He grabs your support limb. You bring that capture near your wrapping hand.

Your hand snatches his wrist/forearm. Push/pull release. Strike the enemy.

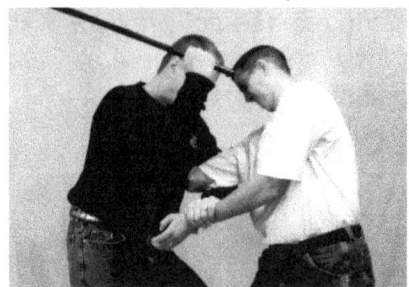

The Beginnings of the Takedowns

We have looked at a large majority of entries into combat scenarios by studying the *Spartan Drill,* the *Chain of the Stick Drill* and the *In the Clutches Of Drill.* There is one more fundamental lesson before beginning the takedowns. Positioning. Your position in regard to his arms, or say to his front, right or left of his arm.

> Your combat scenario set-up is:
>
> Set-up 1: You are in a police, combat, or self-defense situation, whether you are on defense or offense.
>
> Set-up 2: You engage with the *Spartan Drill, The Chain of the Stick Drill* or the *In the Clutches Of Drill* methods, or some similar combat connection.
>
> Set-up 3: Positioning: You will be in three fundamental positions just before the takedown process -
>
> Position 1 - outside his right arm.
> Position 2 - inside his arms.
> Position 3 - outside his left arm.

You will be outside his right arm.   *You will be inside his arms.*   *You will be outside his left arm.*

Stick Controlling and Locking

We next begin looking at takedowns. Many of these grappling captures, in just a second before the takedowns actually occur, may become the so-called "stick lock," if the takedown action is stopped. Or, in the business of law enforcement, corrections and military prisoner capture and control, they might be called "stick control measures."

Many of the combat scenarios you will see on the upcoming pages contain these locks or controls to set up the throw, trip or takedown.

On a personal note, having arrested about a 900 people in 26 years, I am not fond of wrapping up my stick in elaborate and complicated locks in and on people. This ties up my impact weapon, too!. I would prefer to keep it as free as possible. Also, many of the "tap-out" stick locks I have been taught were of little practical value "in the streets," as they say.

Takedown Study Group 1

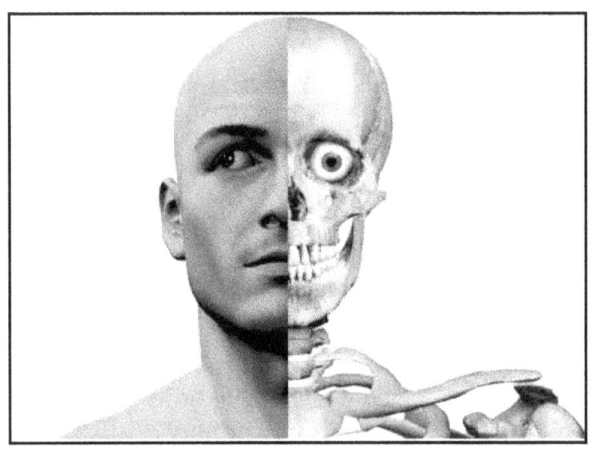

USING THE
HEAD FOR TAKEDOWNS

Injuries to the brain and skull are quite serious and should be left to life and death situations. We need not rehash a basic anatomy class to discuss how important the head is and what a vital nerve, sense and balance center it contains. This is an indictment against violent strikes to (and from) the head, but grappling manipulations do not usually cause such severe damage. The old martial arts, grappling expression is, "where the head goes the body goes" and this is usually a solid and true rule.

In really desperate times, desperate measures may be employed, and striking the head with an impact weapon is most surely going to render "takedown results" if the strike is delivered properly with a proper weapon. Most enforcement agencies simply render all head strikes as taboo and illegal. In some jurisdictions, simply raising a weapon above the head is indicative of a head strike and deemed a violation of policy. High strikes are not practiced or allowed, yet high strikes are not always meant for the head. They could be any downward strikes to an arm to jar a weapon-bearing limb and disarm a weapon.

Removing high strikes is both a major and ignorant disservice to enforcement and security personnel. You consult with your local authorities, rules of engagement and use of force policies about the head strike. And still, in the dark of the darkest night, when you are about to die, and your only option is a head strike to save yourself and live, I hope you have the muscle memory and the wherewithal to strike the head of the enemy, despite official sanction. Do or die.

Needless to say, striking the head is a major takedown method. After striking the head, the next major head related takedown is catching or squeezing the face/head with enough control to twist and turn the head downward into a takedown.

Imagine being this police officer. What is he thinking? Amongst an angry crowd. Excited dog in one hand. A baton raised high in the other. Being filmed from every angle. No matter what he does, he will be deemed wrong by someone.

Head Takedown 1: Any Strike to the Head

Any significant strike to the head, with any significant impact weapon is capable of taking down an opponent. As we progress through all the body parts, please remember this stunning, diminishing strike is always an option.

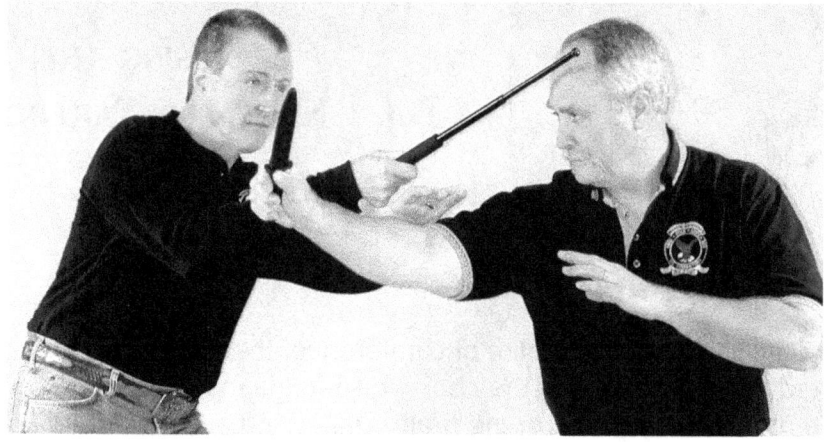

Hit with your stick or even his stick. Here, Tom Pierce threatens me. His real "weapon" at the moment is his big mouth. His stick just supports his threats at the moment. I grab his stick and hit him in the head with it.

Head Takedown 2: The Spital Staircase, Face Vice Takedown

You execute the primary combat scenario steps that lands you outside the arm of a disarmed enemy. He is stunned. You punch the shaft onto his face, reach around the back of his head with your free hand and grab the far end your stick. Pinch the face like a "paper cutter." Pinch anywhere on the face, as it all hurts. Do not pull his head into your torso, instead leave all the action out on your extended support arm, as though you are playing a violin. Pull back and downward. Step away with your inside leg as you do. In this open space, creating a downward, turning spiral takes him down. (the spiral staircase) Just before he hits the ground, let go of your support hand grip and be free from his fall. Once he is on the ground, take the appropriate action to finish the scenario.

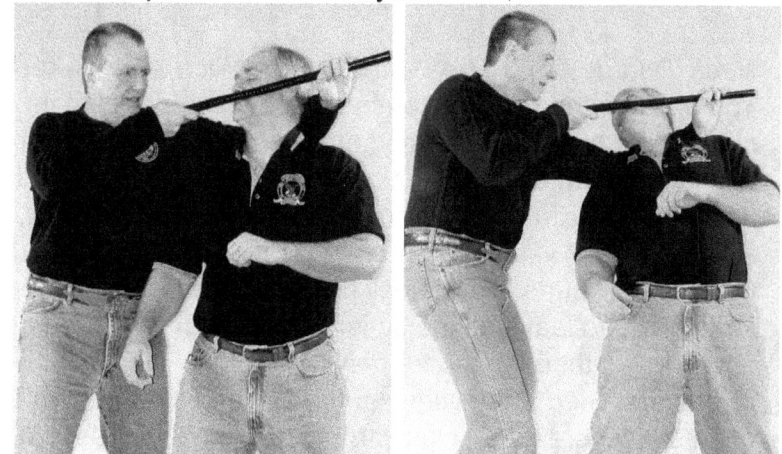

Get the enemy in the painful face vice, step away and twist him down in a spiral, turning with him.

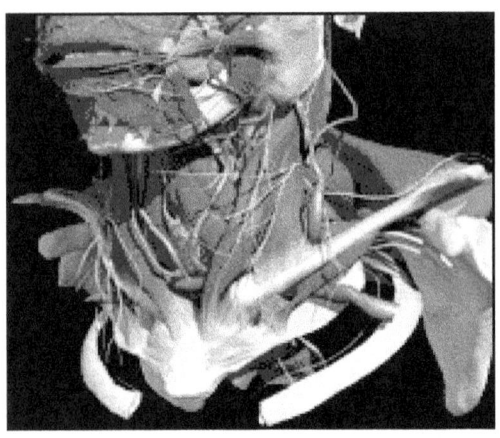

Takedown Study Group 2

Using the Neck for Takedowns

As a target, the neck includes its entire circumference, the front throat area, the two sides of the carotid arteries and the back of the neck which holds the spinal cord and all of its major "electric wiring" into the *computer,* or the brain. Damaging any of these vital spots will injure or even kill an opponent. This segment will look at striking and pinching the neck in vices and chokes that will lead to takedowns. Anytime your stick touches the throat of another, it is a dangerous endeavor. Many enforcement agencies no longer allow any stick contact to the throat of any suspect, and this even confuses justified deadly force issues.

Neck Takedown 1: Strike the Neck Takedown
Any significant strike to the circumference of the neck, with any significant impact weapon is capable of taking down an opponent. A strike to the back of the neck may disrupt the brain. A strike to the carotids may interrupt the blood flow to the brain supported by nerve messages that may knock an enemy off his feet. A strike to the throat may injure or crush his windpipe. Reserve the neck for the most extreme circumstances.

A pommel strike to the throat after blocking a strike, or clearing a single-hand choke.

Neck Takedown 2: The Spiral Stairs Neck Vice Takedown
You execute the primary combat scenario steps that lands you outside the arm of a disarmed enemy. He is stunned. You punch the shaft onto his neck, reach around the back of his head with your free hand and grab the far end your stick. Pinch the neck like a "paper cutter." Pinch anywhere on the neck as it all hurts. Do not pull his head and neck into your torso, instead leave all the action out on your extended support arm, as though you are playing a violin. Pull back and downward. Step away with your inside leg as you do. In this open space, create a downward, turning spiral takes him down. Just before he hits the ground, let go of your support hand grip and be free from his fall. Once on the ground, you take the appropriate action to finish the scenario as the situation calls for it.

The neck/throat in a vice.

Neck Takedown 3: The Rear-Pull Neck Takedown

This maneuver is a basis for many variations. To establish this basis, you execute the primary combat scenary steps that land you outside the arms of an enemy. He is stunned. Step completely behind him, insert your impact weapon across the throat and pull backward and down for a takedown.

Some variations include dropping to your one knee and leaving the other knee up as a weapon. This requires you to step back and away from the enemy as he falls, but not completely away. Leave your knee up. Judge your space. As he falls, position your knee so that his backbone hits your knee. Another target for the knee is the back of the enemy's head. This requires even more space back. Some practitioners can get both a knee to the backbone and a knee to head by moving their knee-high bodies into fast positions.

The simple rear pull is a basis for many variations.

Neck Takedown 4: Fist Ram Choke, Rear Pull Takedown

This is another variation from the rear pull position. As shown in the photo to the right, your fists are in prime position to slip inward and slam the side of the enemy's neck that is essentially two strikes to the carotid arteries. These may further stun and even knock the enemy out. This also affords a tight choke with the shaft on the windpipe and fists on the carotids, providing you pull the head into your chest to tighten all four corners of the choke.

A vicious follow up would be to churn the stick handles back and forth in a rocking motion with your own body torso twisting to add to the dynamic. This violently covers the entire neck. Walk backward as you do this to disrupt his base of balance.

Then you decide when to drop him to the ground. This is a military-style application because of the extent of damage you can do to the enemy.

One variation is to ram the fists into the sides of the neck, since they are in a prime position to do so.

Neck Takedown 5: The Interlocking Choke Takedown

To establish this basis, you execute the primary combat scenario steps that land you outside the arms of a disarmed enemy. He is stunned. Step completely behind him and take the arm and stick positions as shown in the photo to the right and on the next page. This move is taught as a stick choke worldwide, yet without adding improvised hand moves on the carotid and windpipe, it is simply not a choke.

In the photos on this and the next page you will see how I must hook the stick in the bend of my left arm and then I must get my left hand on his right carotid. Then I have to raise my right thumb to squeeze the windpipe. Without these extra hand moves, this would not be a choke. It would simple be a neck restraint as the suspect would have one full carotid supplying blood to the brain and his windpipe open for oxygen.

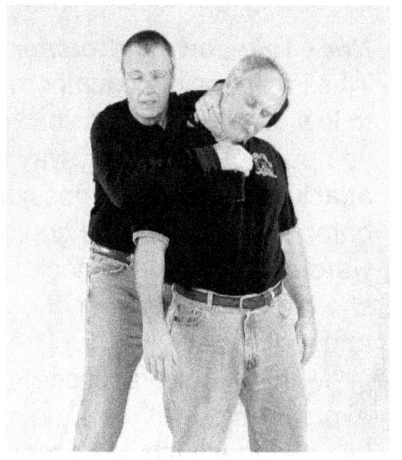

A classic stick choke? It needs more attacks on the neck for a real choke!

Catch the stick in the crook of your arm. Wrap the head. Get the extra attacks on the side of the neck and windpipe. Squeeze and push to stop the blood and wind. Be stepping backward in this process.

When you "choke someone out" while standing, right when they pass out, you suddenly have the full body weight of the opponent to deal with. This could be devastating to your back as you are reflexively prone to capturing and holding that weight. Be prepared to release the subject. Or, since you should be stepping backward anyway to keep the enemy off balance and base, you might step back and slowly lower him to your knee and/or the ground. As he passes out, gravity will take over and put even more pressure on his throat. Warning! Continued choking after pass-out can actually lead to killing your opponent! Who are you fighting and why?

Many martial systems spend a great deal of time mystifying the simple choke-out process. Common sense, practical escapes are good to know, but then these experts bring in all sorts of uncommon escapes and, before you know it, the simple choke becomes this chess game of chain moves and art expertise, when it really should be checkers. Real fighting is more like checkers and less like chess. De-mystify the choke and understand its raw essentials.

Neck Takedown 6: Counter the Tackle Takedown

This is a possible counter to common middle to low tackles. Just as you would if unarmed, you pull your lead leg away from the diving attacker. Both legs, if possible. Grasp your baton in a two-handed grip and shove downward on the tackler's neck. You can try to shove straight down or, if need be, downward and to the sides.

Many of our practitioners worldwide have experimented with this move for more than two decades with great results.

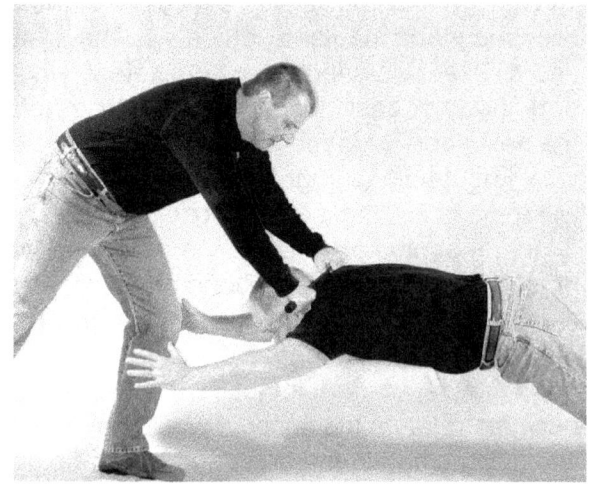

Neck Takedown 7: The X Choke Takedown Series: The Rear

These X-chokes were taught to me at the US Army military police academy in the early 1970s. Due to the level of violence involved with these chokes, they are not taught today to police agencies. They are shown here for historical purposes. The first one is a rear X-choke.

You execute the primary combat scenario steps that land you outside the arm of a disarmed enemy. He is sufficiently stunned. You step to the rear. Put your right hand/stick hand atop his left shoulder and whip the stick shaft down and across the face. Grab the far end with your free left hand and tighten the neck hold. Hand grips into the neck. Push your elbows out to tighten. This squeezes the windpipe and the carotids. Yank this capture forward and back to further disable and confuse him.

Pull him down on the handle side for the takedown. Often you can brace the back of his head against your left knee and really accelerate the pain and the choke.

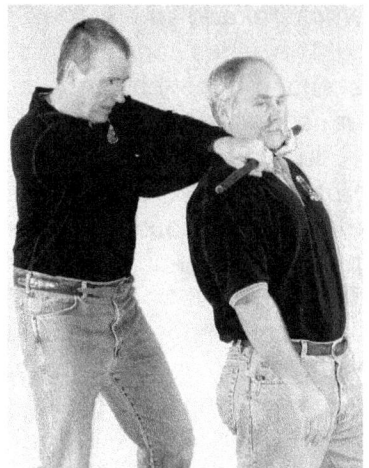

The rear X-choke.

Neck Takedown 8: X Choke: The Front

The series continues. You execute the primary combat scenario steps that lands you inside the arms of a disarmed enemy. He is sufficiently stunned. Put your right hand/stick hand atop his left shoulder and whip the stick shaft down behind his head. Grab the far end with your free left hand and tighten the neck hold. Hand grips into the neck. Push your elbows out to tighten. This squeezes the windpipe and the carotids. Yank this capture forward and back to further disable and confuse him.

Observers dream of attacking the eyes and groin of the choker, but the first instinct from the stunned enemy once caught in this X-vice is to grab at the stick, his neck, and your forearms.

Pull him down on the handle side for the takedown. Often you can brace the front of his head or face against your left knee or shin and really accelerate the pain and the choke.

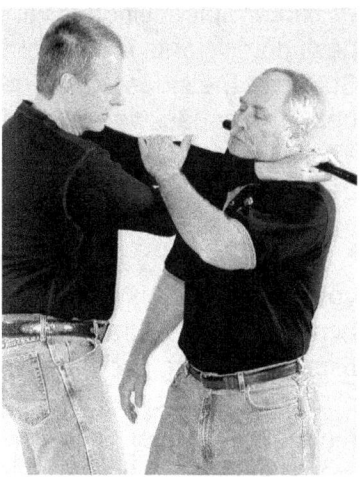

The Frontal X-choke.

Neck Takedown 9: X Choke: The Sides

The series continues. You execute the primary combat scenario steps that lands you inside the arms of a disarmed enemy. He is sufficiently stunned.

You feed the stick and grab in the manner shown in the photo to the right. Bring your hands, wrists and forearms together. Shake his head and neck violently to further confuse him and haul him down. As he hits the ground, let the lower hand grab on to the stick go.

It is important that you know this capture, if done in a certain angle and manner, could cause grave injury and possibly even death. You must practice with slow experimental caution and execute in the field with this same caution.

It is often easier to start these chokes with a reverse grip stick. Experiment with this on the right and left sides.

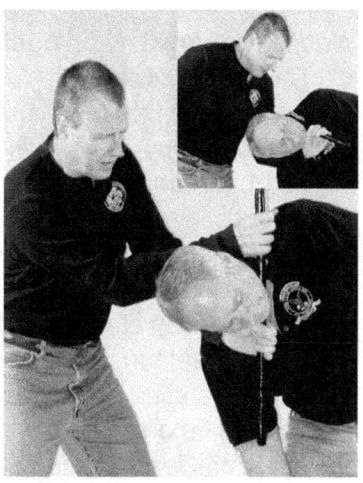

You may place the high stick tip into your armpit.

Neck Takedown 10: The Supported Chokes

The series continues. You execute the primary combat scenario steps that land you just outside the arms of a disarmed enemy. He is sufficiently stunned.

In this next set of two options, perhaps seeing them on film or in person better explains the movements. I include them here because they are classic stick choke attacks that practitioners should experience, but they might not make the final cut in everyone's favorite survival profile.

You punch the shaft forward against the neck. You reach with your empty hand around the back of the enemy's head and once looping the head, grab the stick near where your original hand grip is, but on the pommel space. This original grip slides down the stick and positions the stick end in either of these two support positions:

    a) stick behind your neck.
    b) stick buried into your the armpit.(better).

Stick braced across the back of my neck. Note that the cover hand holds the free arm.

The armpit position is very secure compared to the neck position. His and your height matters in this decision. Squeeze the sides and front of his neck as you splay your upper body backward. Your free hand covers his other arm. You will note in the photos that the armpit catch creates a better V-shape on the neck and produces a better choke-like capture.

Crank to disrupt, then hammer fist his nose, then wipe your arm down his free arm to control. Step backward to get him off balance. Crank the choke again, multiple hammer fists to the nose and face. Then wipe your arm down his free arm to regain control of that arm. When you decide to take him down, do so in a downward spiral as your inside leg steps away and back. If the stick is in your armpit position, let go of your other grip as the enemy falls down, maintaining the stick in your armpit. Spiral takedowns.

Stick braced inside my armpit.

Neck Takedown 11: Wheel Throw on Back of the Neck

The series continues. You execute the primary combat scenario steps that land you just outside the arms of a disarmed enemy. He is sufficiently stunned.

With a 2-hand grip you strike a hooking blow into the opponent's lower stomach causing him to double over. If he has doubled over far enough, you can continue with this move. If you wish, you can simply push down on the back of his neck as in Neck Takedown 6. But, if his head is not low, this wheel throw is a good option.

Move the stick to the back of his neck, and try to maintain his inside arm up as high as possible. Turn the head inward, and clear your legs from his path of falling.

Belly hit! Pictures continue on next page....

Loop the head with the shaft of the stick. Strike and catch the back of the neck. Keep his arms up with good stick positioning. Spin him down. Get your legs out of the way.

Neck Takedown 12: Stomach Choke, Gravest Extreme

The neck series continues. You execute the primary combat scenario steps that land you just inside the arms of a disarmed enemy. He is sufficiently stunned. Strike him in the stomach, and he bends over. Get your stick across his throat as you get the back of his head on your stomach. If you get both ends of your stick in the bends of your arm and your hands on his shoulders? WARNING! This will usually kill the enemy, beyond just choking him and it will crush his windpipe. This is a lethal force move reserved for kill-or-be-killed action. Study this, and do not to do it by accident.

 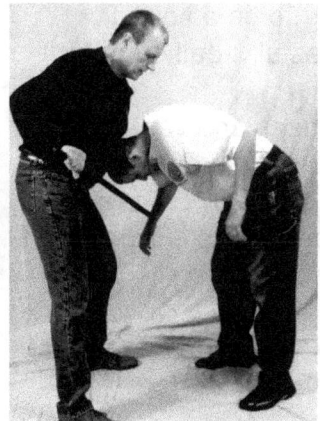 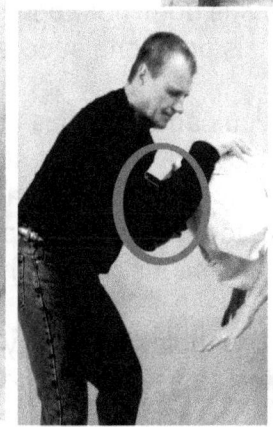 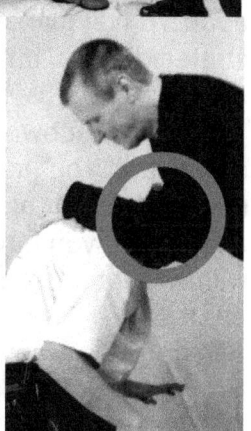

Takedown Study Group 3

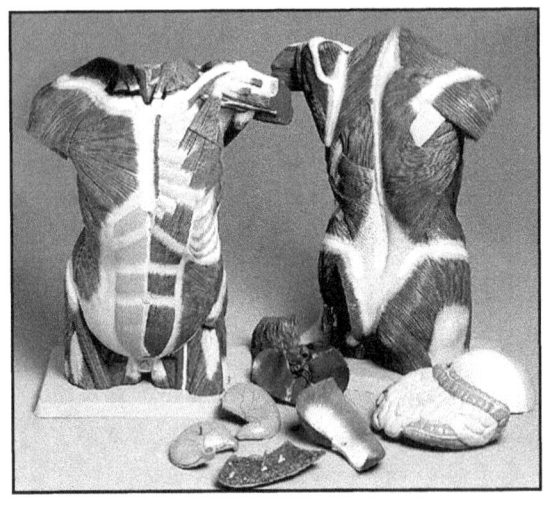

Using the Torso for Grappling and Takedowns

For the purpose of this study, the torso includes the part of the body from the top of the shoulders on down to the pelvis. Not the neck. Not the arms. Not the legs.

Torso Takedown 1: The Clavicle Pull Down

This is a maneuver that is a basis for many variations. To establish this basis, you execute the primary combat scenario steps that land you outside the arms of a disarmed enemy. He is stunned.

You get the shaft of your impact weapon to the far side of his head and crash it downward and slightly inward. In severe circumstances, you break his collarbone. Otherwise the strike is more of a pull. Drop your own body weight, bending at the knees, to create the force to take him down.

If you attack the shoulder closest to you, your stick could slide off.

Torso Takedown 2: The Chest Pull Back

This maneuver is a basis for many variations. To establish this basis, you execute the primary combat scenario steps that land you outside the arms of a disarmed enemy.

This is a grappling, containment move. Typically enforcement agents may use this to break up a fight and pull people apart. At first glance this does not appear to be a sound strategic move, but it is one done successfully, daily around the world.

For a takedown you can buckle his knees and yank downward. You must quickly yank the stick away (sideways may be best) or he may instinctively grab the stick to prevent a fall. Also, you have to make sure the person he is fighting with will not charge and hit the man in the face as you pin his arms.

There are many psychologies at play when you interrupt a fight, and such are the subject of another book.

Torso Takedown 3: Rear Pelvis Pull

This is a maneuver that is a basis for many variations. To establish this basis, you execute the primary combat scenario steps that land you outside the arms of a disarmed enemy. This is also an old police and military escort technique from a simple rear approach of a suspect or an enemy.

As the photo to the right displays, with a center grip, you get the stick between his legs and turn it horizontally. You push on the upper body. You can walk him off or take him down.

The controversy begins when "experts" tell you that your palm should be palm-up or palm-down. They will argue that the suspect will yank the stick up for palm up or push the stick down versus the palm-down grip. Both are sensible counters.

I have done this in police work, and the suspects usually swing their arms wildly anyway. I prefer the palm-down grip.

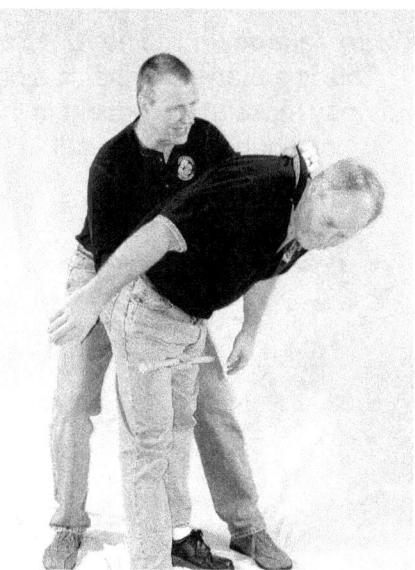

The rear groin pull.

Torso Takedown 4: Front Pelvis Pull

This is a maneuver that is a basis for many variations. To establish this basis, you execute the primary combat scenario steps that land you inside the arms of a disarmed enemy. This is also an old police and military escort technique from a simple front approach of a suspect or an enemy.

As the photos below display, with a center grip, you get the stick between his legs. You can use this as a forearm strike to the groin. You turn the stick horizontally. You push on the upper body. You can walk him off or take him down.

I have done this move also in police work, and it is surprising how much strength a stick bearer has in these positions.

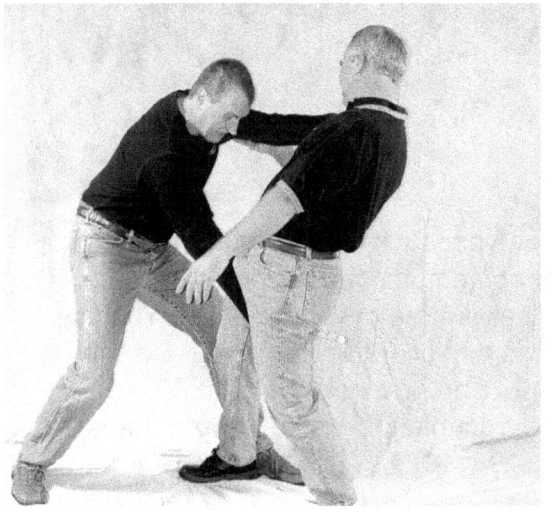 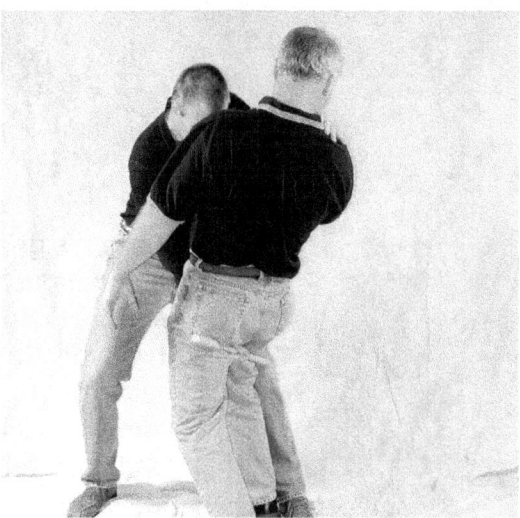

You are on the inside of a stunned opponent. You can strike the groin with your forearm and pass the stick between his legs. Twist the stick. Pull on the bottom, push the head or neck on the top. As an option, you might drop the horizontal stick down behind the knees.

Torso Takedown 5: The Riot Stick Tip Strike to the Stomach

This is a maneuver that is a basis for many variations. Using the classic riot stick position, you may strike the opponent in the solar plexus with enough thrusting force that the opponent may well double over and fall.

The classic riot stick thrust to the solar plexus can produce a takedown.

Torso Takedown 6: The Stomach Throw

This is a maneuver that is a basis for many variations. An enemy, sometimes bigger, faster or stronger than you gets you in a wrestling match over your baton, in more or less of a "port arms" position. He charges, and you are moving backward, about to fall anyway. Or you may be tripping. This is the classic Japanese *tomoe nage* throw or, so nicknamed by many since the late 1960s, as the *Captain Kirk* throw from Star Trek because he tossed more than a few Klingons and other assorted aliens around with it.

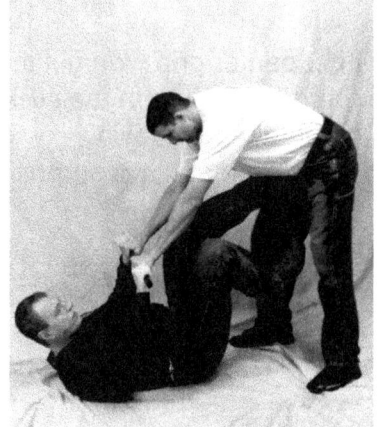
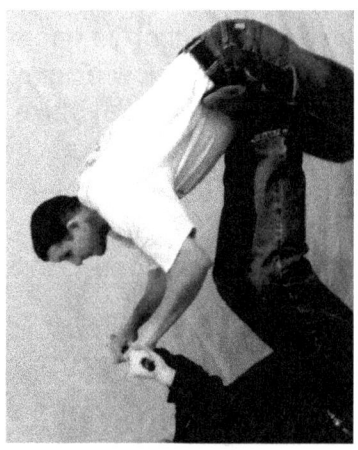

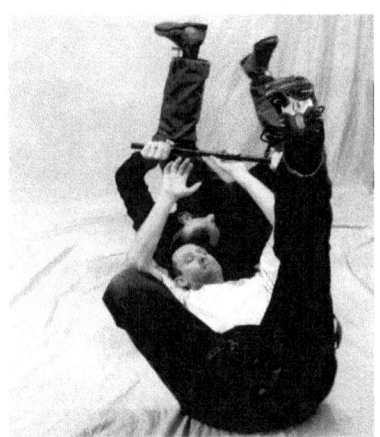

Using his grip upon your stick (or long gun), with his forward, pushing momentum, you fall/pull and place your feet into the pelvis area of your attacker. You use this dynamic to toss him over the top of you. He will fall between 10, 11, 12, 1 or 2 o'clock over you. You must yank the stick free from his grip as he realizes he is landing harshly.

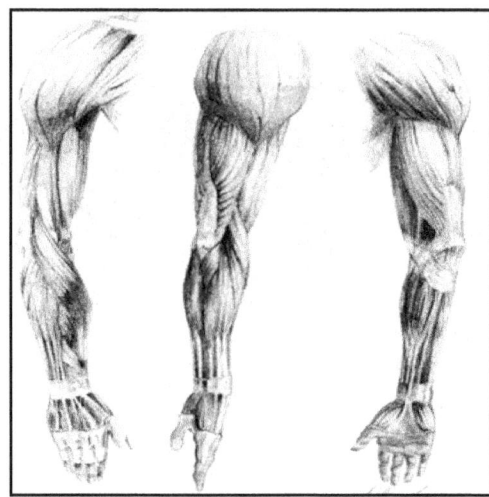

Takedown Study Group 4

USING THE ARMS FOR GRAPPLING AND TAKEDOWNS

This section studies the use of the arm as a takedown tool. These range from the upper arm and armpit, the triceps, the biceps, elbow, forearm, wrist and hand. This also may involve, at times, using the torso as a support. Even the neck may be used for support. There are numerous, intricate and, at times, complicated moves here that many martial arts practitioners play with, that I will not do here in this very practical, tactical book.

Arm Takedown 1: The Biceps Lever

This maneuver is a basis for many variations. To establish this basis, you execute the primary combat scenario steps that land you inside the arms of a disarmed, stunned enemy.

You grasp the arms and twist it out, twisting the palm outward. This requires a trained grip. You place the stick between the arm and the torso. You use the stick as a crowbar and shove it against the biceps, twisting the arm, bringing his arm back and down.

Some practitioners will step on the opponent's foot to prevent him from stepping back and catching his balance.

After some physical practice with this, you will understand what the best position of the twisting arm will generate the torque for the takedown. You can also add trips with the legs.

Arm Takedown 2: The Triceps Lever

This may start off as a common "old school" police or security escort. This is a maneuver that is a basis for many variations. To establish this basis, you execute the primary combat scenario steps that land you outside the arms of a disarmed, stunned enemy. You grasp the arm and twist it out, twisting the palm outward. You place the stick between the arm and the torso. You use the stick as a crowbar and shove it against the bicep, twisting the arm, bringing him forward and down. Some practitioners will sweep his inside leg to prevent him from stepping forward and catching his balance. After some physical practice with this, you will understand what the best position of the twisting arm will generate the torque for the takedown.

I have used this escort position numerous times. It works if you use the elbow area and not the shoulder. If he resists, you can try this push, trip and turn or go to another position.

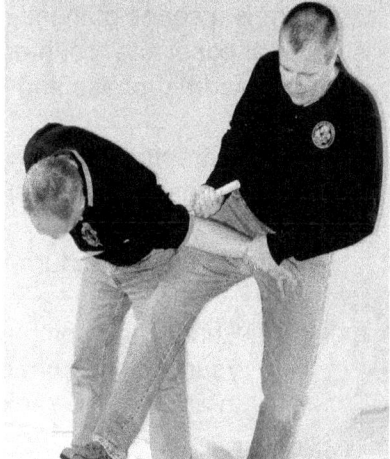

Arm Takedown 3: The Bent Arm Takedown

This maneuver is a basis for many variations. To establish this basis, you execute the primary combat scenario steps that land you inside the arms of a disarmed, stunned enemy.

You get the proper bend in the arm and keep it bent toward his body. You run the stick parallel with the forearm and you turn the stick backward. Drive his elbow forward until it is pointing forward and until he tips over backward. Do not lose the bend in the arm. Step off to get the best position to facilitate this move.

Note in the photograph how my top hand grabs both the top of my stick as well as secures his wrist. There are several different ways to do this, such as running your forearm parallel to his forearm and hooking the stick handle on his wrist. Let go of that grip as he falls.

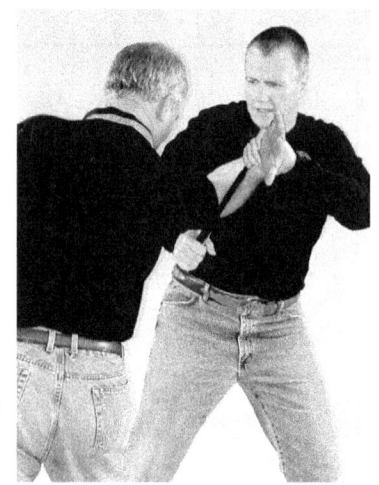

Arm Takedown 4: The Arm to Neck Bridge Takedown

This maneuver is a basis for many variations. To establish this basis, you execute the primary combat scenario steps that land you inside the arms of a disarmed, stunned enemy.

You get the proper bend in the arm and keep it bent. You run the stick tip under his arm and across to his neck. Use his arm as an axis and your stick as a crowbar. Grab his wrist and crank it back and downward. Do not lose the bend in the arm. Step off to get the best position to facilitate this move.

Extract the stick when you know he is tipped over and falling.

Arm Takedown 5: The Snake Killer Takedown Series

This maneuver is a basis for many variations. To establish this basis, you execute the primary combat scenario steps that land you inside the arms of a disarmed, stunned enemy. This is the *Snake Killer* because it is a counter to having your impact weapon wrapped or "snaked" - as in the enemy doing a snake disarm.

Here is a high-yield move. As he wraps and snakes your stick, you catch the wrist, NOT the hand, of the enemy. You turn it outward and down. You "churn" the stick on his forearm bones. This creates a high yield takedown that feels and works far better than it looks. Step off to get the best position to facilitate this move. Extract the stick when you know he is tipped over and falling.

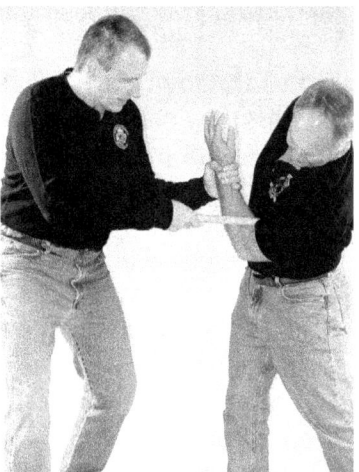

There is a series of these moves that practitioners should also study that constitute "counters to a snake disarm" that fall into the early-phase, mid-phase and late phase counters.

> 1: Extract the stick by yanking it out of the snake wrap. (This is not a takedown, just an escape in this set.)
>
> 2: Do the above turn-out takedown.
>
> 3: Punch your pommel downward toward his hip, trapping his hand. You reach to grab your stick tip from behind his back. You grab and turn that tip forward. This catches him in an arm bar. Yank him forward and down.

Arm Takedown 6: The Straight Side Arm Bar

This maneuver is a basis for many variations. To establish this basis, you execute the primary combat scenario steps that land you outside the arms of a disarmed, stunned enemy. With your empty hand, you run clockwise over the enemy's limb until you forearms are connected as shown in the photo to the right.

You run your stick over his arm, hitting on or about the elbow with the shaft. You pull him downward with this elbow catch. Try not to draw him into your legs. Step back and away from him and try to drop him onto his chest.

If he manages to bend his arm, rolling his elbow up, you can proceed to the next capture - the Rear Arm Bar Hammerlock.

Arm Takedown 7: The Rear Arm Bar Hammerlock

This maneuver is a basis for many variations. To establish this basis, you execute the primary combat scenario steps that land you outside the arms of a disarmed, stunned enemy. With your empty hand you run clockwise over the enemy's limb. You bend his arm inward until his hand touches his spinal cord area. Keep your hand hooked near his elbow NOT on his triceps or certainly not his shoulder. The elbow is the key to control.

To escape pain in his shoulder he bends over. You can bring him back upward with your other hand. You can do this arm bar with empty hand or your stick hand. Or you can let him bend over to a fall.

To finish a takedown, you can use these two examples:
- buckle his leg with your knee, and drive him face down.
- let him bend and turn him in a downward spiral.

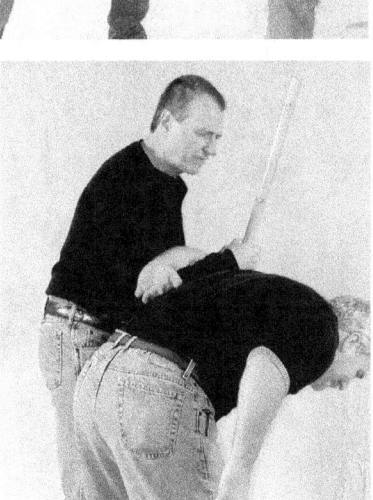

The Choke Option of the Rear Arm Bar Hammer Lock: If you hook with your weapon side as shown in the center of this page, and the rear arm bar is secure, feed your stick in the manner shown in these photos. In the center photo move you can even push the pommel with your left hand to better position the shaft of the stick. My training partner Barnhart here is very broad and stout, but it is possible to get your stick almost vertical with other people depending on their size. Big or small, this choke, once seized, is painful and immediate.

Arm Takedown 8: The Branch Down Bent Arm Bar

This maneuver is a basis for many variations. To establish this basis, you execute the primary combat scenario steps that land you inside the arms of a disarmed, stunned enemy.

In this case, you are confronted by an impact weapon wielding mad man, who is more interested in yelling at you than swinging a stick. His voice and face is his weapon at the moment. His stick is up and presented.

You grab his stick mid-height, mid-shaft, hit his head with his stick, then pull his stick tip down. Hit him with your stick as you insert his stick tip between his hand and body. Get the stick as close to the elbow as possible for leverage. Grab the end. Pull. Knee. Pull down.

If the enemy lets go of his stick early on in this? You now have disarmed him. If he lets go halfway, you can still get the catch.

You grab his stick mid-height, mid-shaft, hit his head with his stick, then pull his stick tip down. Hit him with your stick as you insert his stick tip between his hand and body. Get the stick mid-shaft.

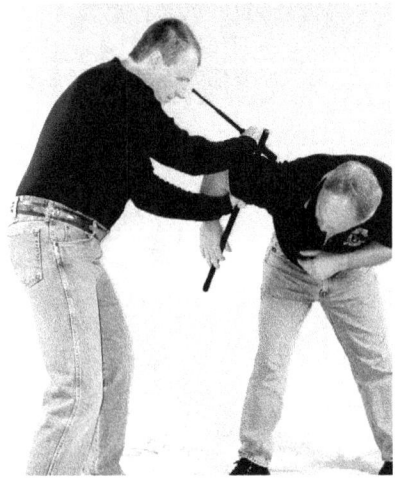 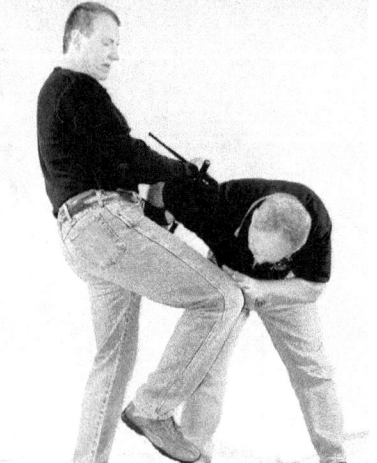 

Hit his head with his stick, then pull his stick tip down. Hit him with your stick as you insert his stick tip between his hand and body. Get the stick as close to the elbow as possible for leverage. Grab the end. Pull. Knee. Pull down.

Arm Takedown 9: The Reverse Arm Bar

This is the opposite motion of the previous *Branch Down Takedown 8*. This is a maneuver that is a basis for many variations. To establish this basis, You execute the primary combat scenario steps that land you inside the arms of a disarmed, stunned enemy.

You feed his stick tip over the top of his arm and continue this feeding circle until his stick tip is under his elbow. You turn toward the same direction he is facing, until you are both facing the same direction. If you step over to face him and turn his elbow from that 3 o'clock position to an 11 or 10 o'clock position, you will tip him over backward for a takedown.

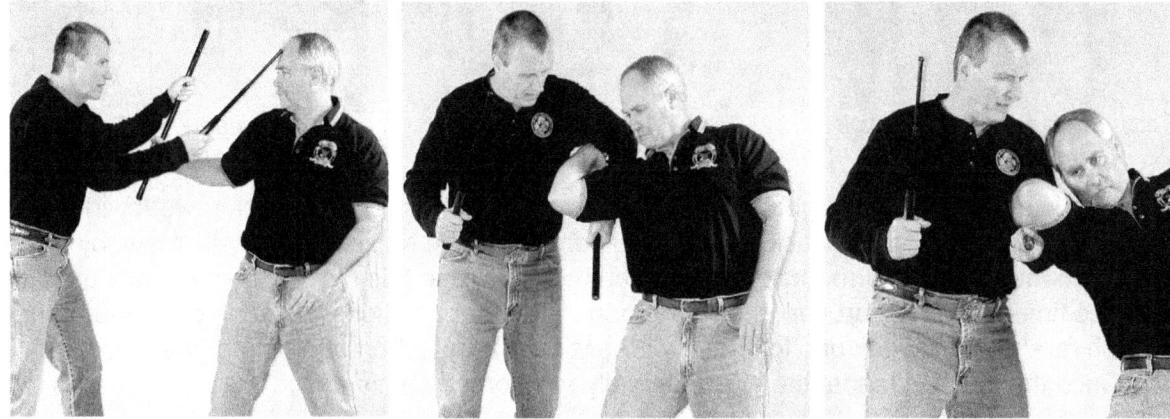

Arm Takedown 10: The Branch Down Push or Punch Catch

This maneuver is a basis for many variations. To establish this basis, you execute the primary combat scenario steps that land you outside the arms of a disarmed, stunned enemy. This attack could be a high push, or a wounded, slower punch.

You block the push or punch with your impact weapon. You see it linger long enough to insert the stick above and over his forearm. You hook your thumb over his wrist and, as he pulls back and away (a common motion), you ride that back and turn his elbow up. Get the stick near his elbow for leverage. Grab the end of your stick and turn forward for a takedown. Walk and move with the turn to power the takedown. As a police officer I have manhandled and arrested suspects with this very move. I have done it versus a hard push and versus someone that had grabbed my uniform shirt. I have never executed it versus a full-speed punch, though I have seen some very slow punches from drunks and untrained people.

Remember that the person will likely retract his arm and you can follow it back for this catch. You follow this motion in with your thrusting stick.

He pushes or punches. You block. Open up your thumb from the baton and hook his wrist. Make his arm bend, or just follow the very common arm retraction.

Catch the far end of your stick behind his elbow.

Pull out a bit, then you turn him forward and down.

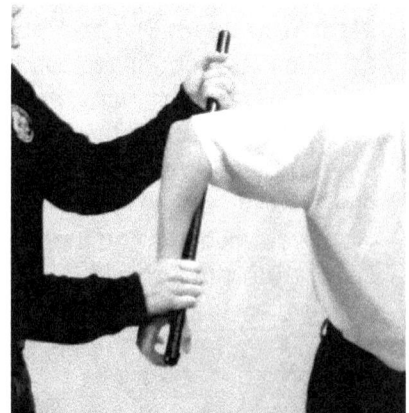
Use his elbow, NOT the upper arm and NOT the shoulder.

This move is one of the mistaught and misunderstood escorts taught to policing worldwide. The common mistake is to put the stick on the shoulder, where the subject can easily resist by using all his torso muscles. The arm must be significantly bent to use this move. I myself have used it numerous times as an escort grip, then transitioned to other positions. Use this open thumb hooking catch rarely. It is a weak grip to use any other time except when attempting this catch and carry. Once the catch is made, the grip is actually reenforced by his arm.

Arm Takedown 11: The Bent Arm Pulldown

This maneuver is a basis for many variations. To establish this basis, You execute the primary combat scenario steps that land you outside the arms of a disarmed, stunned enemy. In this situation he is throwing a high hook punch. You pass it in front of you. You strike the chin with an uppercut stick strike. Get the stun! Then with a two-handed grip get the crook of the arm. Pull down in the direction of the captured elbow.

He hook punches.

You strike the head. He is stunned.

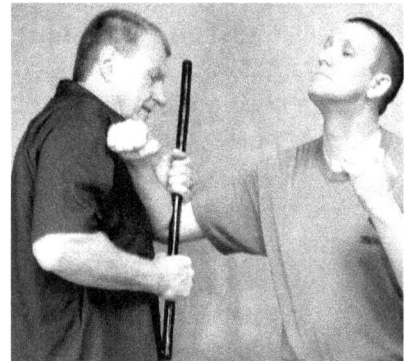
"Catch the crook."

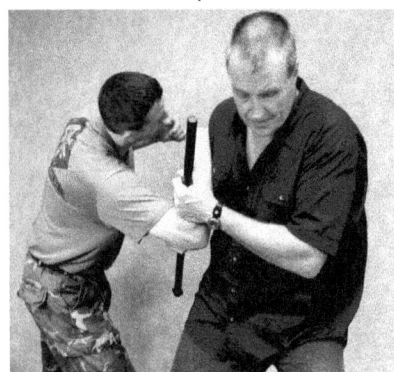
Aim the elbow at 8 o'clock. Step and pull in that angle

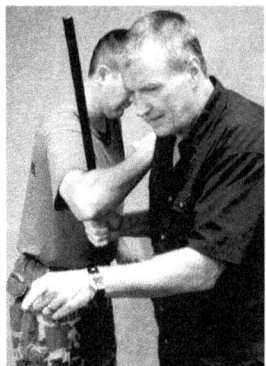
Best angle.

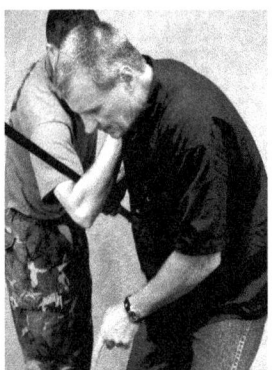
This angle is over his leg. Not good.

This angle too high.

Takedown Study Group 5

Using the Legs for Grappling and Takedowns

This section studies the use of the legs as a takedown tool. These range from the pelvis, the upper leg, thigh, knee, lower leg, ankle and foot.

Leg Takedown 1: The Big Catch and Rollover
This is a generic, leg-catch maneuver that is a basis for many variations. To establish this basis, you execute the primary combat scenario steps that land you inside or outside the arms of a disarmed, stunned enemy. You scoop and catch the enemy's medium height or high kick. Five ways to catch a medium or high kick are:

Left arm hook arm up catch.

Right arm hook arm up catch.

Left arm hook arm down catch.

Right arm hook arm down catch.

Center catch.

The opponent threatens. He kicks.

You catch this kick.

With this grab. You roll his leg over and pull down and back.

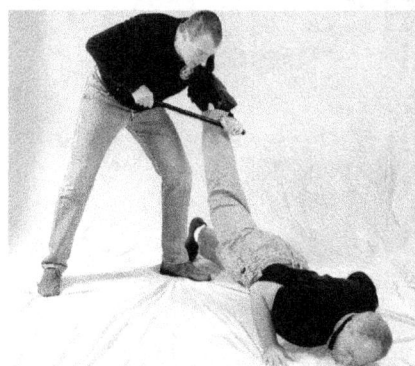
You pull him face down.

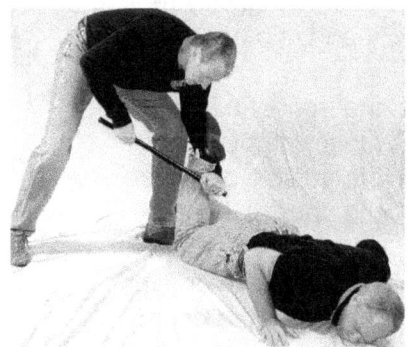
If he continues, you smash his knee.

Leg Takedown 2: The Pelvis Pull Takedown

This maneuver is a basis for many variations. The enemy may be stunned from the first events of a fight. This is a situation where you have been knocked down by the slow or stunned opponent, and you land to the side of him.

You insert the stick between his legs from behind his legs. You brace/lay the stick across his pelvis, grab the far end of your stick with your support hand, getting a 2 hand grip. Then you pull him down, breaking his balance across his pelvis line.

In the photo to the right, Barnhart poses to show where the stick is about to be positioned. Right onto the pelvis is best.

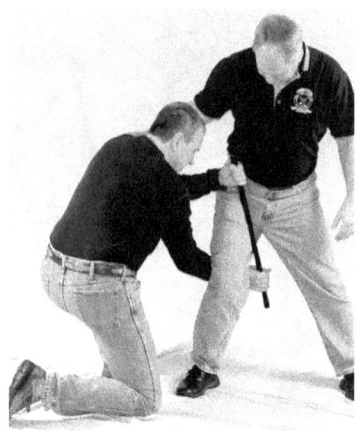

Leg Takedown 3: The Thigh Shove Takedown

This maneuver is a basis for many variations. To establish this basis, you execute the primary combat scenario steps that land you inside the arms of a disarmed and stunned enemy. As the photo demonstrates, it is a sound idea to step on the foot. Trap the foot. Next, you insert the stick between the legs and then turn your grip hand palm up, laying the stick across the top of the thigh. The far end of the stick braced against the back of one leg and shoved across the thigh of the another leg. You shove the stick down. Experts will attempt to support this move with an outer twisting wrist action on the caught hand. Twisting and shoving, you charge forward.

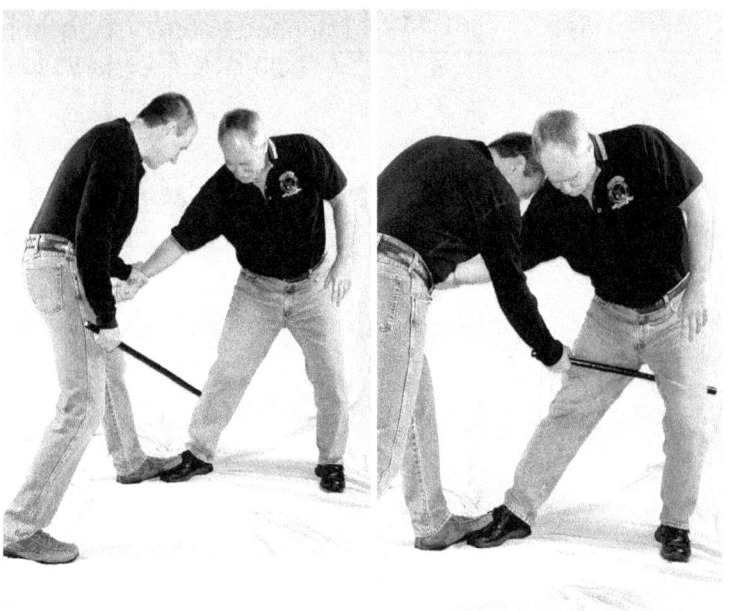

Leg Takedown 4: The Knee Push Takedown Series

This maneuver is a basis for many variations. The enemy may be stunned from the first events of a fight. This is a situation where you have been knocked down by the slow or stunned enemy. You quickly respond with a 2-hand grip and shove against:

    Example 1: the front of a knee
    Example 2: the inside of a knee
    Example 3: the outside of a knee
    Example 4: the back of the knee

You may have to use two or even three of these angles as the enemy tries to slip away.

Leg Takedown 5: The Knee X-Pull

This maneuver is a basis for many variations. To establish this basis, you execute the primary combat scenario steps that land you outside of a disarmed, stunned enemy.

Barnhart poses in the photo to the right, to allow a demonstration of this grip. You get an X cross-forearm grip right on the knee or just below the knee. You tighten the X by driving your elbows outward.

You lift, pull and turn on this capture, and it is extremely painful. You turn the enemy right off of his feet, over and down. Extract the stick when possible.

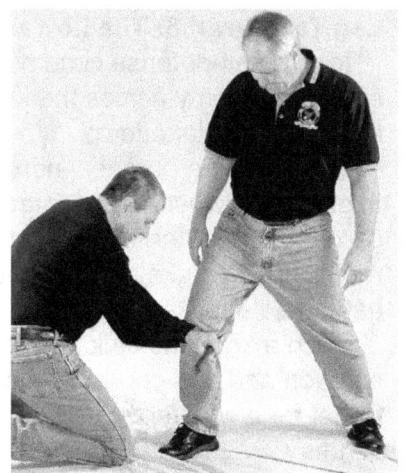

Leg Takedown 6: The Ankle X-Pull

This maneuver is a basis for many variations. To establish this basis, you execute the primary combat scenario steps that land you outside of a disarmed, stunned enemy.

Barnhart poses in the photo to the right, to allow a demonstration of this grip. You get a X cross-forearm grip right on the ankle or just above the ankle. You tighten the X by driving your elbows outward.

You lift, pull, and outer twist-turn on this capture, and it is extremely painful. You turn the enemy right off of his feet, over and down. Extract the stick when possible.

Leg Takedown 7: The Fireman's Pole

This maneuver is a basis for many variations. To establish this basis, you execute the primary combat scenario steps that land you inside the arms of a stunned enemy.

Controlling his arms, you bear hug, clinch the enemy, slip down (ergo the classic "fire pole") and run your stick behind his legs, stopping at the bend of his knees.

You pull in at the bend and push against his torso, for a push/pull takedown.

Pull your stick out when you know he is committed to the fall, otherwise you may loose your stick under his legs.

Leg Takedown 8: The Lower Leg Strike (such as kneecap)

In dire self-defense circumstances, a soldier or citizen may strike the enemy across the kneecap. This is a high-yield, fight-finisher and a takedown.

The common peroneal nerve strike across the side of the thigh (about where the outside seam of a pant leg runs on the center of the outer thigh) is the law enforcement and security preferred leg strike. It hurts, but usually does not stop a semi-dedicated attacker.

Such a kneecap strike could be challenged by police administration and civil courts. A knee strike is an unconventional target for law enforcement, and as such will undergo great scrutiny.

If you really need to stop a subject and put him down in a desperate situation, a shattered kneecap is one major way. A similar strike to the shin may also be effective.

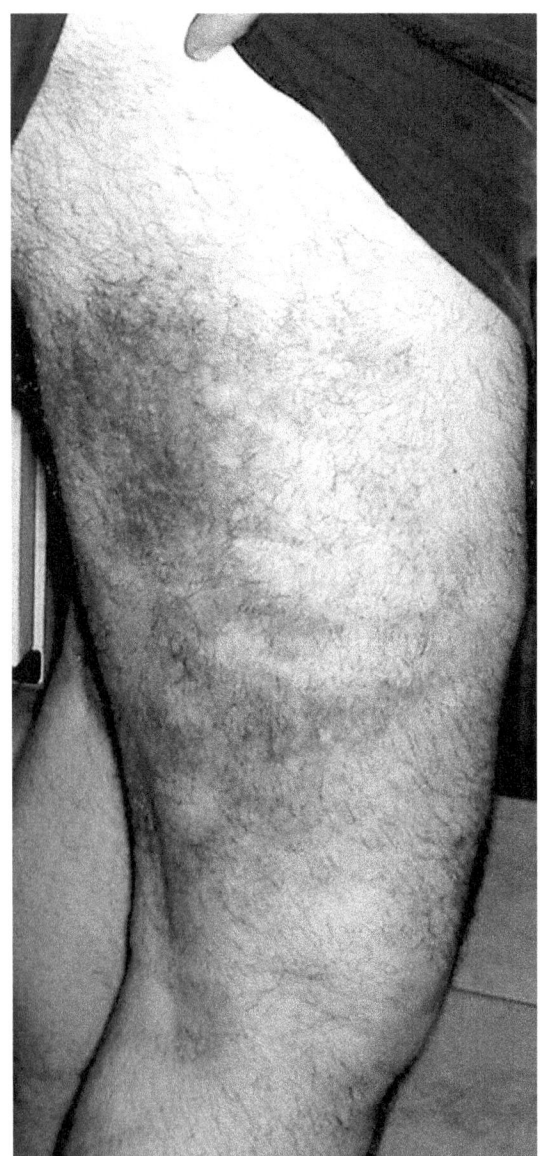

Lifelong Texas martial artist Dean Goldade displays the bruises he's received from yet another common day of Filipino stick fighting. He, like myself and countless others who have "hard-stick-fought," have never fallen to the ground after having been struck on the outer thigh with an impact weapon. A researcher would have a hard time finding officers with these successful nerve strikes as takedowns.

Bruised? Yes. Hurt? Yes. A takedown? Probably not. At some point, worldwide law enforcement trainers will have to accept the fact this commonly taught nerve strike, while perhaps a law suit preventive, will not save an officer's life.

(Warning, this type of injury with extensive bruising from stick fighting can result in blood clots, stroke and even death. We do not suggest or recommend sparring with force significant enough to cause bruising of this sort. Should bruising result from stick practice, we recommend you consult a physician.)

Takedown Study Group 6

Walking Cane and Umbrella "Hook Handle" Takedowns

Mark Shuey demonstrates a handle catch.

A walking cane and/or umbrella may have several handle designs. Flat, curved or knobby. The curved handle offers the best shape for grappling. The shaft of a cane or a specially prepared and hardened umbrella can be used for blocking and grappling in the same manner that is shown on the prior pages.

But, the hook offers unique capabilities to the common walking stick. The hook, if wide enough, may capture a neck, an arm or a leg. And those captures, in conjunction with other body part captures and moves, may produce a takedown or throw.

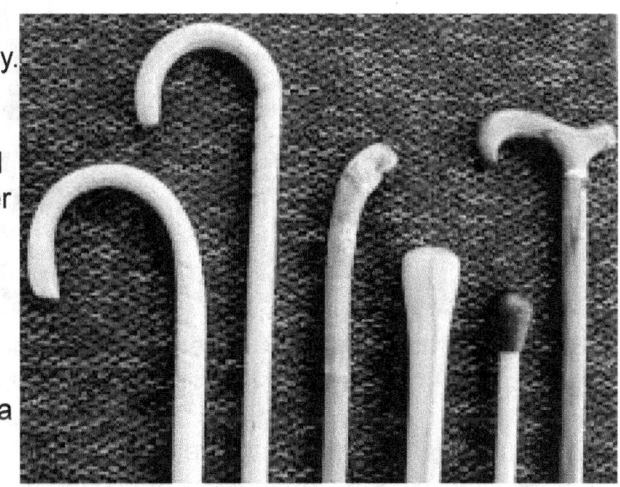

The flat handle, the knobby handle, or even the popular curved handle may be used as a striking tool.

Mark Shuey demonstrates a hip throw after a neck hook capture.

The handle hook catches:

- the neck, its front, side, and back.
- the arm, its front, side, and back.
- the leg, its front, side, and back.

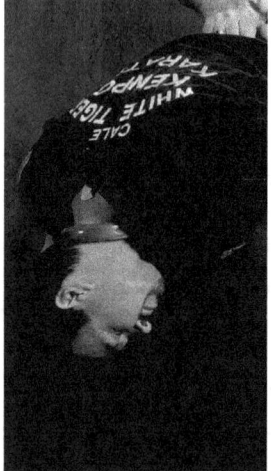 

Once you hook the neck, violent pulls or pushes on the neck can really manipulate the head and then the body. You can hook the lower leg and cause tripping. You can hook the arm and, with that arm capture, along with another body capture and movement, cause a takedown. Mark Shuey manufactures and sells a variety of canes with hooks large enough for these limb and neck tactics. Plus, he makes some cane handle hooks that have sharp enough handle ends that the tip can also be used as a painful weapon in counter-attacks.

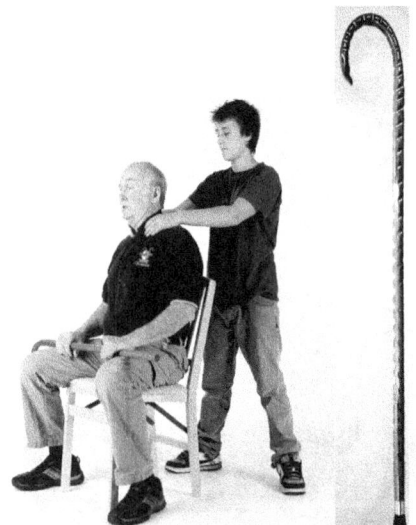  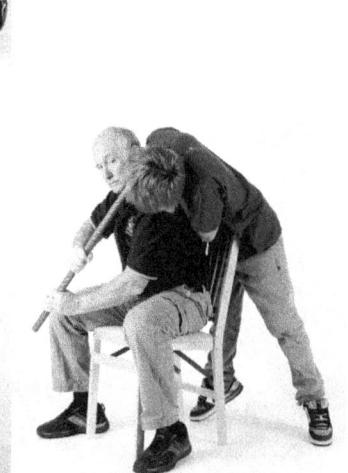

Mark Shuey hooks the neck of a rear attacker while he is seated. Once the neck is hooked, the attacker can be violently jerked around and forced into various positions.

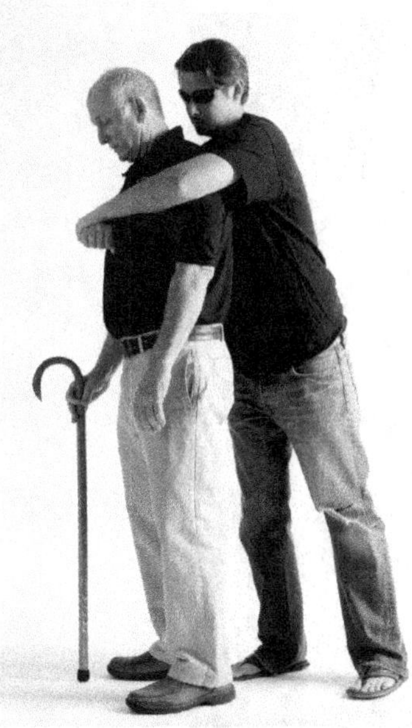

Here, the sharp tip at the end of the handle can also be an effective tool against an attacker. Mark Shuey custom builds these tips into many of his canes. The grooves on the cane are important for gripping. Common canes do not offer these grooves, as they are not designed for self-defense. Shuey's canes are designed to solve this problem.

Contact Mark for training and canes at:
Cane Masters
7050 W. Palmetto Park Rd. Suite 15-255
Boca Raton, FL 33433
866-642-1898 www.canemasters.com

Brave pensioner Doris Thiele from the U.K. demonstrates how she fought off a serial burglar with a walking stick.

A Japanese ad for the unbreakable umbrella.

You can find advertisements for these sturdy and "unbreakable" umbrellas all over the internet. The tip, shaft and handle striking potential is the same as all the tactics shown in this book and as demonstrated with the ever common, stick shape. It is the curved handle of the cane or umbrella that offers the unique grappling options.

Beat and treat these umbrellas in training as you would a wooden or plastic stick. The folded material of the upper umbrellas will of course cushion the strike somewhat, but not too much.

Impact Weapon: The Takedown Skeleton List

If a practitioner knows how to execute the following takedowns with an impact weapon, he or she knows a great deal about the subject. This foundational study can be used for experimenting, understanding, creating and collecting even more.

Basic Head Takedowns
1: Head Takedown 1 - Any significant strike to the head.
2: Head Takedown 2 - Face Vice Takedown.

Basic Neck Takedowns
3: Neck Takedown 1 - Strike to the Neck Takedown.
4: Neck Takedown 2 - Neck Vice Takedown.
5: Neck Takedown 3 - Rear Neck Pull Takedown.
6: Neck Takedown 4 - Fist Ram Choke Rear Pull Takedown.
7: Neck Takedown 5 - The Interlocking Choke Takedown.
8: Neck Takedown 6 - Counter the Tackle Takedown.
9: Neck Takedown 7 - The X Choke Takedown Series: The Rear.
10: Neck Takedown 8 - The X Choke Takedown Series: The Front.
11: Neck Takedown 9 - The X Choke Takedown Series: The Sides.
12: Neck Takedown 10 - The Supported Choke on the Back of the Neck.
13: Neck Takedown 11 - Neck Takedown Wheel Throw on the Back of the Neck.
14: Neck Takedown 12 - The Stomach Choke.

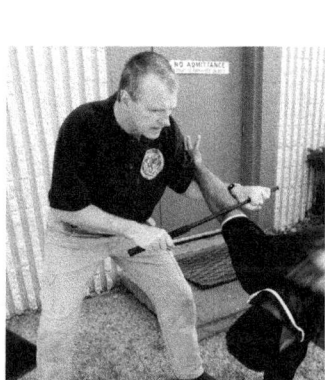

Basic Torso Takedowns
15: Torso Takedown 1 - The Clavicle Pull Down.
16: Torso Takedown 2 - The Chest Pull Back.
17: Torso Takedown 3 - The Rear Pelvis Pull.
18: Torso Takedown 4 - The Front Pelvis Pull.
19: Torso Takedown 5 - The Riot Stick Tip Strike to Stomach.
20: Torso Takedown 6 - The Stomach Throw.

Basic Arm Takedowns
21: Arm Takedown 1 - The Biceps Lever.
22: Arm Takedown 2 - The Triceps Lever.
23: Arm Takedown 3 - The Bent Arm Takedown.
24: Arm Takedown 4 - Arm to Neck Bridge.
25: Arm Takedown 5 - The Snake Killer Takedown Series.
26: Arm Takedown 6 - The Straight Side Arm Bar.
27: Arm Takedown 7 - The Rear Arm Bar Hammerlock.
28: Arm Takedown 8 - The Branch Down Bent Arm Bar.
29: Arm Takedown 9 - The Reverse Arm Bar.
30: Arm Takedown 10 - The Branch Down Push or Punch Catch.
31: Arm Takedown 11 - Bent Arm Takedown.

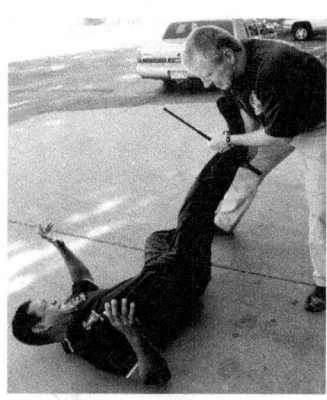

Basic Leg Takedowns
  32: Leg Takedown 1 - The Big Catch and Roll-over.
  33: Leg Takedown 2 - The Pelvis Pull Takedown.
  34: Leg Takedown 3 - The Thigh Shove Takedown.
  35: Leg Takedown 4 - The Knee Push Takedown Series.
  36: Leg Takedown 5 - The Knee X-Pull.
  37: Leg Takedown 6 - The Ankle X-Pull.
  38: Leg Takedown 7 - The Fireman's Pole.
  39: Leg Takedown 8 - The Lower Leg Strike (such as kneecap).

Basic Walking Cane Hook Handle Takedowns
  40+: Hook arms-wrists, neck, legs-ankles, clothing.

With this knowledge as a foundation, continue to develop your personal list of takedowns in these various categories. Remember, it is critical to understand the "Diminished Fighter Theory" and that you usually have to out-speed, distract, stun, wound or injure the opponent to execute takedowns.

Grappling Trick- Grow and row one end.
From a 2-hand grip, feed the pommel on one end. Use the long side to "row" past people in a crowd. We train this using a gauntlet of people in two lines. A smaller feed can reach around and strike the base of the skull.

CHAPTER 17: IMPACT WEAPONS ON THE GROUND

A survival rule of thumb bears the question, "Can I do this on the ground?" We have thus far asked you to try every draw, retention, strike, block and grappling standing, kneeling. seated and on the ground. By this chapter you should be exposed to the "bottom-side of stick fighting."

We must remember that by the common martial term, "ground" in the real world means many surfaces other than the semi-trampoline flooring of the Octagon and mats. The "bottom" is actually the indoors and outdoors of rural, suburban and urban areas. Rather then me list all of the bottoms from gravel, asphalt, cement to tiles, stairs, carpets and wood, just think about all of them. In my decades as a patrolman and detective I have fought people on many of them, from sliding down a muddy hill in the rain, gravel picnic grounds, sidewalks to industrial carpets and even atop and under furniture.

Since this is a book *not* dedicated to wrestling, we will ask you to review all that we have recorded and all we will overview in this chapter and get on down and experiment.

The problem geography

A thorough "ground" fighting course deals with:

  Seated in a "chair."

  Knee high
    fighting those above.
    fighting those equal.
    fighting those below.

  Down
    on your back.
    on your right side.
    on your left side.

  Unarmed through various weapons.

Basic Unarmed and Armed Ground Maneuvers "While Holding" a Stick

Grounded Number 1: The Ready Position
- do unarmed or with weapons sheathed, holstered or mounted.
- do while drawing a stick.
- do with weapons presented.

Grounded Number 2: The Shrimp, or Hip Escape
- do unarmed or with weapons sheathed, holstered or mounted.
- do while drawing a stick.
- do with weapons presented.

Grounded Number 3: The Shoulder Walk
- do unarmed or with weapons sheathed, holstered or mounted.
- do while drawing a stick.
- do with weapons presented.

Grounded Number 4: The Bucking Bridge
- do unarmed or with weapons sheathed, holstered or mounted.
- do while drawing a stick.
- do with weapons presented.

Grounded Number 5: The Fishtails (Head and Torso)
- do unarmed or with weapons sheathed, holstered or mounted.
- do while drawing a stick.
- do with weapons presented.

Grounded Number 6: The Sit-Up
- do unarmed or with weapons sheathed, holstered or mounted
- do while drawing a knife, a stick, a pistol, a rifle
- do with weapons presented

Grounded Number 7: Guard Rotation
- do unarmed or with weapons sheathed, holstered or mounted.
- do while drawing a stick.
- do with weapons presented.

Grounded Number 8: The Common Rollover
- do unarmed or with weapons sheathed, holstered or mounted.
- do while drawing a stick.
- do with weapons presented.

Grounded Number 9: The Scissor Kick Rollover
- do unarmed or with weapons sheathed, holstered or mounted.
- do while drawing a stick
- do with weapons presented.

Grounded Number 10: Side Rotation, The Curly Shuffle
- do unarmed or with weapons sheathed, holstered or mounted.
- do while drawing a knife, a stick, a pistol, a rifle.
- do with weapons presented.

Grounded Number 11: Back Pivot Rotation
- do unarmed or with weapons sheathed, holstered or mounted.
- do while drawing a knife, a stick, a pistol, a rifle.
- do with weapons presented.

Grounded Number 12: Hip Pivot Rotation
- do unarmed or with weapons sheathed, holstered or mounted.
- do while drawing a knife, a stick, a pistol, a rifle.
- do with weapons presented.

Ground fighting is an enormous subject because it includes almost all standing material also, just done on the floor with a different gyroscope. Again we show this important book For much more hand, stick, knife and gun ground maneuvers, get Hock's book.

Ground weapon retention sample scenario

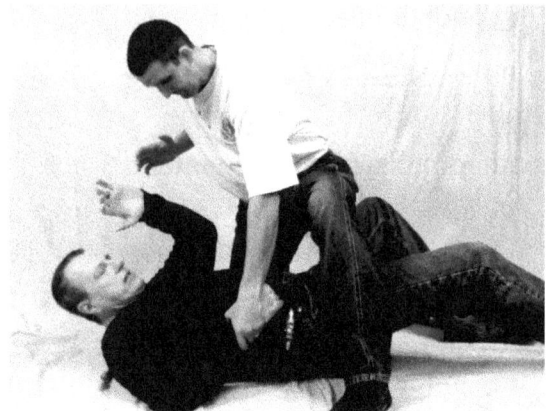 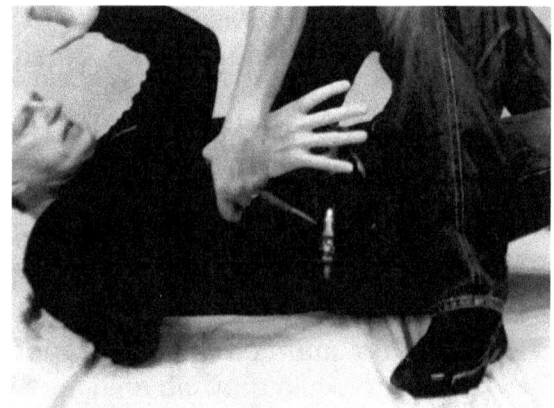

He grabs the limb of your attempted quick draw. You let go of your weapon and remove both your and his hands from your weapon area.

 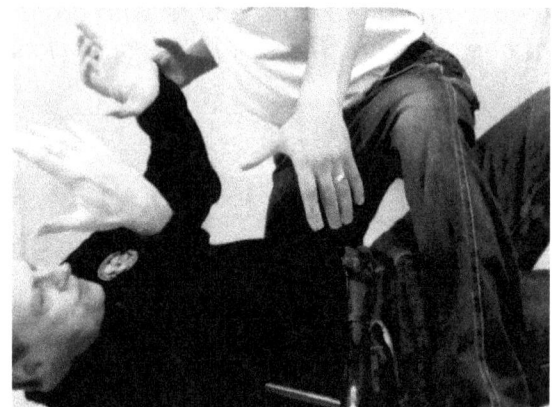

Get any number of hand releases.

If the opponent is "high" enough on you, you might try this leg hook and toss. Then draw.

Ground stress quick draw sample - closed expandable baton - Review

You are taken down. Fight and work to a good position to draw your expandable baton, and strike the enemy in key places with the weapon closed or open. When the opponent is sufficiently stunned, escape or control him as the situation dictates.

You are knocked down.

He comes in. You shoulder walk back. Kick to gain space, and/or draw.

Use the closed baton as a striking weapon. Then open it.

Counter a ground choke sample scenario

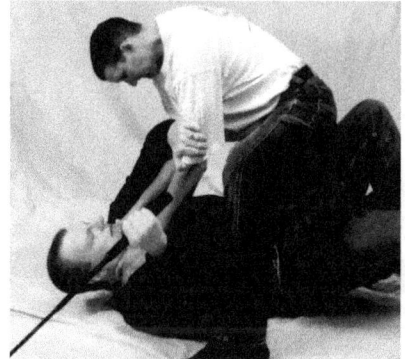

He is choking you. You feel incredible pressure on you windpipe. You cannot waste time.

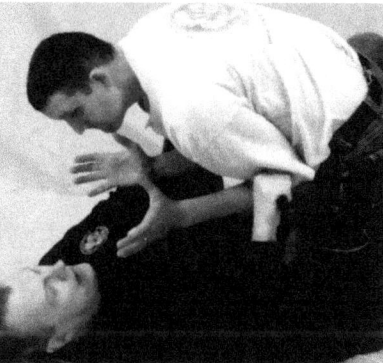

You must release this pressure first. You strike downward at the bend of his arms.

With your airway free, you begin to pummel him with your hands, elbows and stick.

A ground striking sample scenario

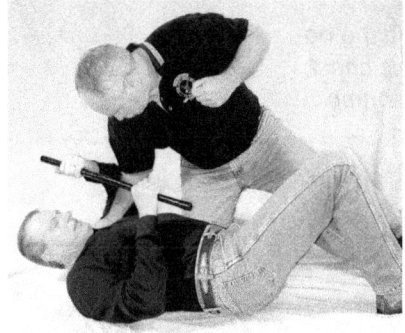

You're down. Use your 2 hand grip to block his pounding.

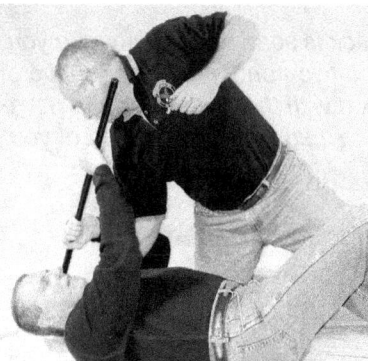

Use your 2 hand grip to strike back.

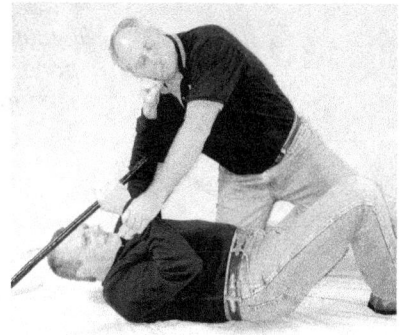

Hand strike back, too.

A ground striking "trenching the skull" sample scenario

If you can, if you need a distraction, or if the head is available as a target trench the skull. Trenching the skull with the pommel is both a painful and bloody experience. As you might imagine, the pommel raked and rubbed anywhere on the face and head would be a good move.

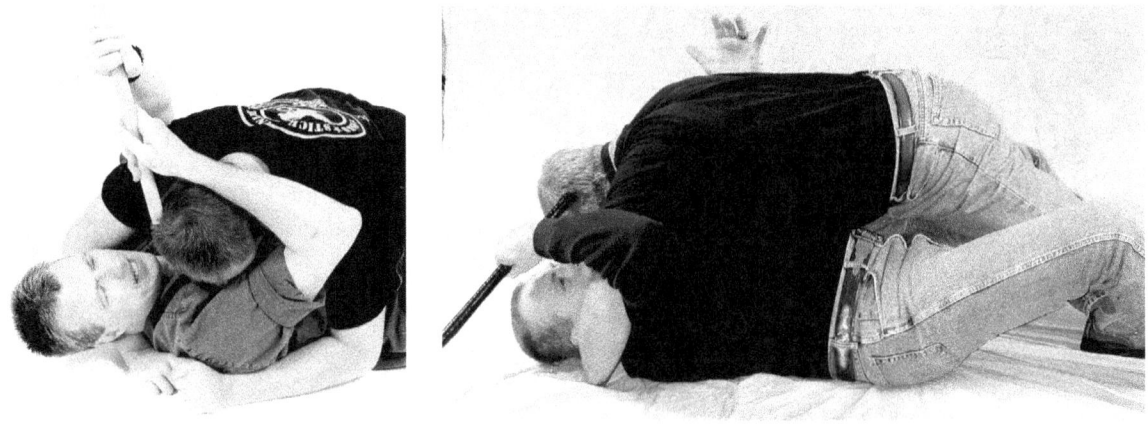

If your stick is sandwiched between you and the opponent, a two-hand grip power shove on his chest, coupled with the classic ground power "shrimp might shove a foe off of you.

Impact weapon ground fight sample - Slider 1- The Driller
Your weapon is out. You are charged and taken down. He lands on top of you, and your stick is pinned somewhere between both your chests. Slide the stick out. Hit him several times in key places, and/or stab the stick into his rib cage. Slip your hand up to the center grip, and drill the stick into his ribs. Accompany this with a powerful hip heist and "body-curve shrimp."

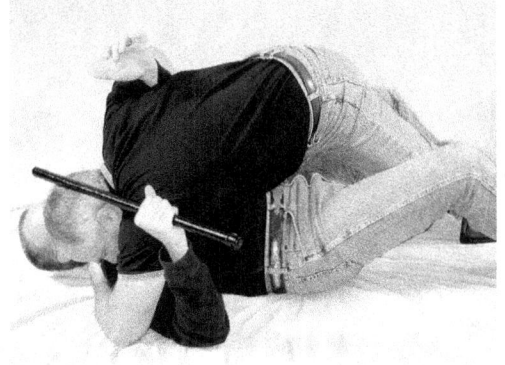 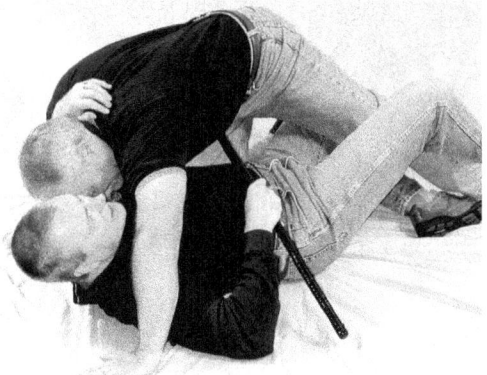

Slide the stick out and brace the tip on the ribs of the enemy.

Impact weapon ground fight sample - Slider 2 - The Choke
Your weapon is out. You are charged and taken down. He lands on top of you, and your stick is pinned somewhere between both your chests, but your stick grip hand is above his arm on your stick grip side. Slide the stick out.

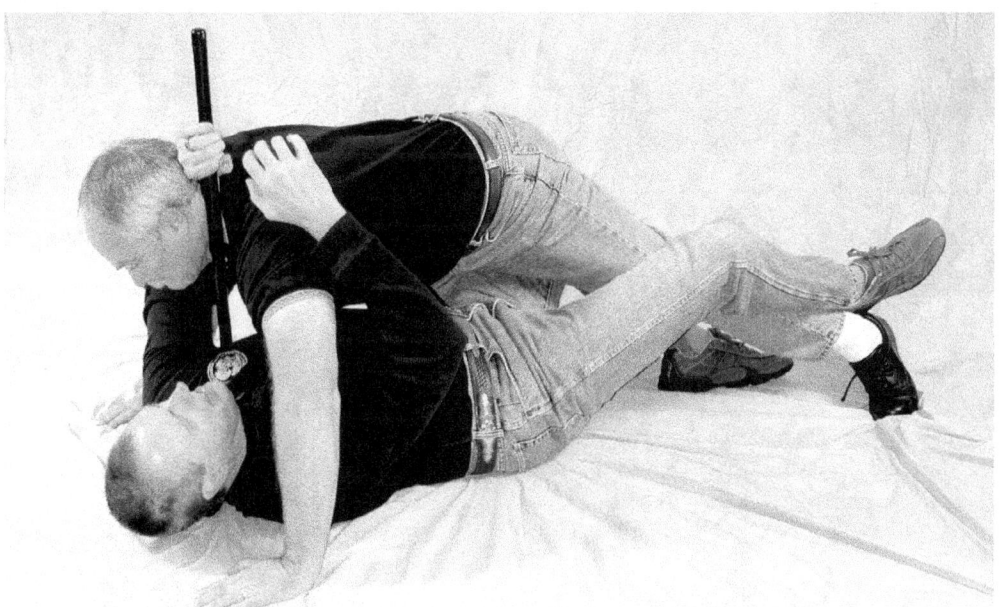

Get the far end of your stick into your arm pit. Pinch it in place. Then grab the end out of the stick with this same limb. Squeeze the sides of his neck, and yank him off to the stick side to escape, or continue the choke from a side position or if you can continue the rollover, and maintain this choke from topside. Kick your legs out to the side. Be heavy.

Impact weapon ground fight sample - The Topside Turn

Your weapon is out. You charged, and he is taken down. You land on top of him, and your stick is the center of a push/pull wrestling match. This is his "bench press" power position. Instead of pushing or pulling against this power, turn the stick left or turn it right. He loses his power position. Then take the top end and force it down over his face (once again the center lock position) and to his throat. Cap off the other end with your palm, and use your stick for a choke.

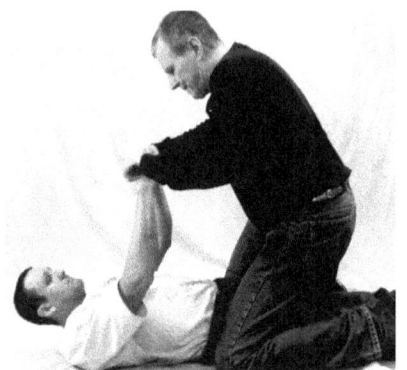

It's a push/pull bench press stand-off!

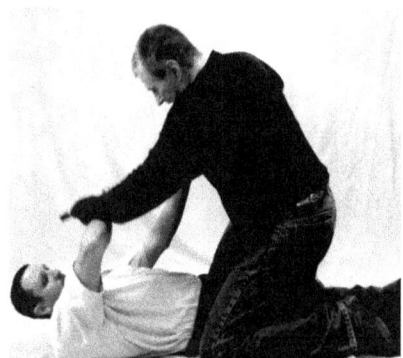

Make a turn. Here is a turn.

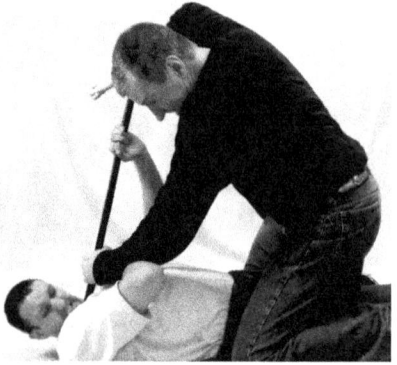

Force the tip down on his throat. Cap off the rear with your palm.

Impact weapon ground fight sample - The Knuckle Dragger

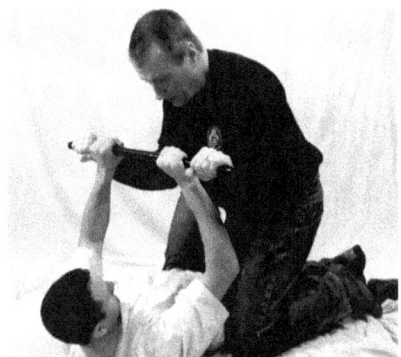

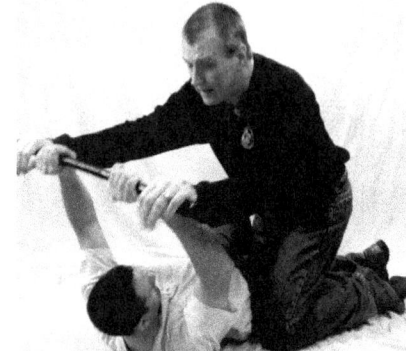

Put your elbows in, and push up and over. Ram the knuckles on the ground. You might "put the grapes in" as in grape-vining the legs.

Drag his knuckles across the ground to get a grip release.

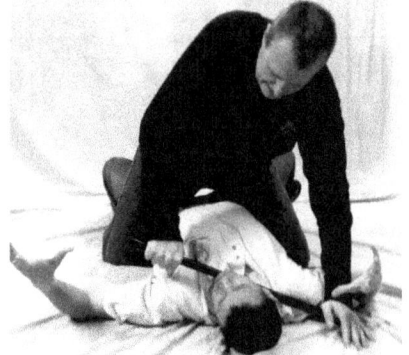

Once free, lay the stick across the neck and choke.

Impact weapon ground review - The Shield Stress Quick Draw

This is a sample scenario for when you are knocked down and about to be jumped. You quick draw your impact weapon and create a shield as you get back to your feet.

Barnhart is sucker punched down.

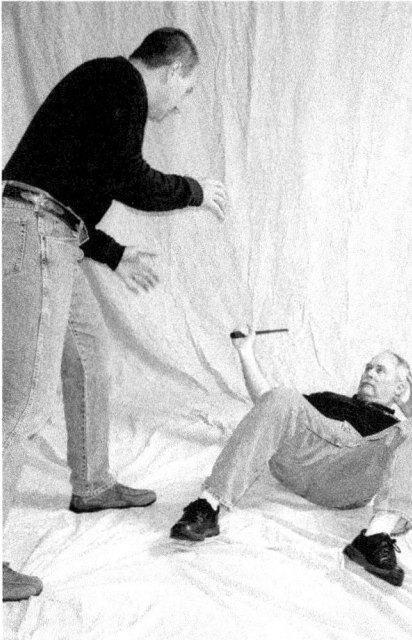
Barnhart draws his impact weapon.

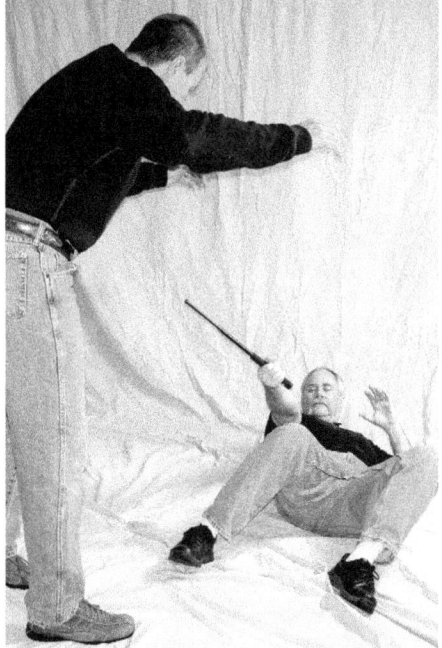
He swings as soon as possible.

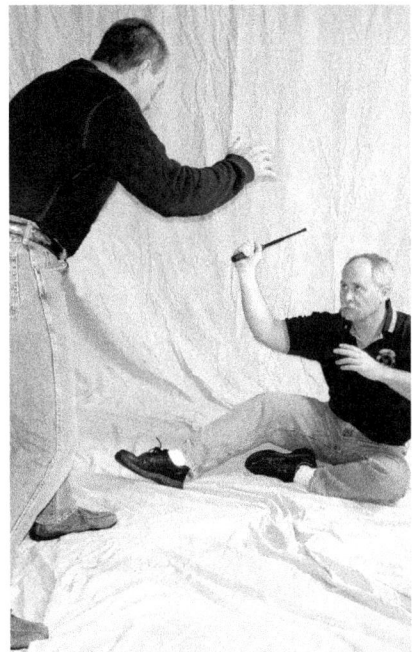
He continues to swing as he...

...gets up. He creates a shield.

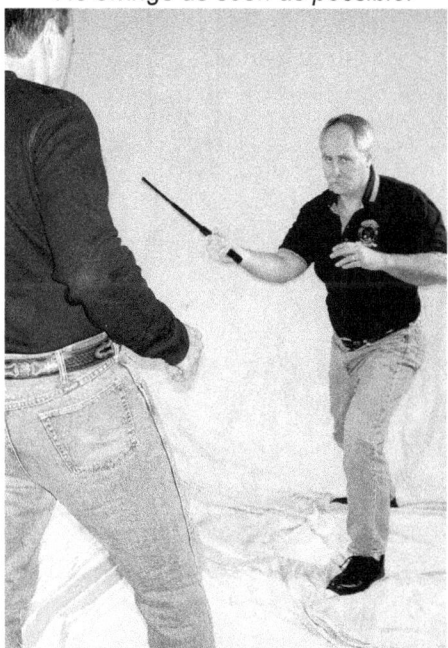
He gets back to his feet.

The Ground Shield

1: The enemy strikes you down.
2: You gain some space with a "shoulder walk" and/or grounded thrust kicks.
3: You draw your baton.
4: You immediately start swinging at the enemy.
5: This creates a shield, keeping him away from you as you get up.

Chapter 18: Test Requirements

Force Necessary Stick: Level 1 Test

Six hours experience working on these Level 1 materials. Time and grade in similar themed systems also may count upon approval.

Knowledge and Understanding
* A working and conversational knowledge of the *Who, What, Where, When, How and Why* Questions concept.

* A working and conversational knowledge of the *"Who-Stick"* question.
* A working and conversational knowledge of the *Stop 6*.
* A working and conversational knowledge of *Stop 1* of the *Stop 6*, and all the pre-fight and pre-crime skills it entails.

* A working and conversational knowledge of the Introduction to impact weapons.
* A working and conversational knowledge on selecting your "stick."
* A working and conversational knowledge on the legal and moral issues of stick carry and use.

Physical Accomplishments
Level 1 Scenarios

* Perform any 10 impact weapon, stress, *Stop 1*, quick draw combat scenarios.
    - which will include the Level 1 of the Stalking Drill.
    - which will include the Level 1 Stick Ambush, Dodge and Evasion Drill.

* The Ready Stance, from nothing to bladed, while holding.
* The Pendulum Step while holding a stick, working around the Combat Clock.

Any Local Instructor Additions.
  * *Current and subject to change   Test fee is $100.*

FORCE NECESSARY STICK: LEVEL 2 TEST

Eight more hours experience working on these Level 2 materials. Time and grade in similar themed systems also may count upon approval.

Knowledge and Understanding
* A working and conversational knowledge of all the *Stop 2* commonalities.
* A working and conversational knowledge of the *What* question, as it relates to batons.
* A working and conversational knowledge of *Stop 2* as it relates to impact weapon fights.

Physical Accomplishments
* The V and Inverted V footwork while holding a stick, on the Combat Clock.
    - while holding a one hand grip.
    - while holding a two-hand grip.
    - add the "turn-ins."

* Demonstrate the Tangler basic releases from *Stop 2* grips and catches on your stick carry site and stick drawn catches on your hand/wrist or forearm.
    - 2 releases from belt line grabs.
    - 2 releases from single hand grabs on your stick.
    - 2 releases from double hand grabs on your stick.
    - 2 releases from a double-hand grab, one on your stick, the other on your wrist.

* Demonstrate the Level 2 Stick Tangler Exercise
    2 stress draws vs. an unarmed attack.
    2 stress draws vs. a stick attack.
    2 stress draws vs. a knife attack.
    2 stress draws vs. a pistol rush.
    2 attacks vs. an unarmed attacker, stick drawn.
    2 attacks vs. a stick attacker, stick drawn.
    2 attacks vs. a knife attacker, stick drawn.
    2 attacks vs. a pistol pulling attacker, stick drawn.

Any Local Instructor Additions.
Current and subject to change test fee is $100.

FORCE NECESSARY STICK: LEVEL 3 TEST

Ten more hours experience working on these Level 3 materials. Time and grade in similar themed systems also may count upon approval.

Knowledge and Understanding
  * A working and conversational knowledge of all the *Stop 3* commonalities.
  * A working and conversational knowledge of the *Where* question, as it relates to batons.
  * A working and conversational knowledge of *Stop 3* as it relates to impact weapon fights.

Physical Accomplishments
  * The Shuffle Foot footwork while holding a stick, on the Combat Clock.
    - while holding a one hand grip.
    - while holding a two-hand grip.

  * Demonstrate stick blocks
    * single hand blocks.
    * double hand grip blocks.
    * counters to common blocks.

  * Demonstrate the Stop 3 Forearm collisions
    - explain and demo the 4 Ps, pinning, passing, pulling, and pushing.
    - explain and demo the outside invasion series
      * show the 4 basics.
      * show 2 advanced applications
      * demonstrate and explain the block, pass and pin drill
      *   - versus unarmed, show six shoves and stress quick draws.
      *   - versus knife, show 4 half-beat inserts with takedowns.
      *   - versus stick, show 4 half-beat inserts with takedowns.

  * Demonstrate the Level 3 Stalk and Mad Rush exercise where the trainer stops his mad rush invasion on the forearm collision.
      * show 1 scenario with a stick quick draw and strike, maybe a takedown.
      * show 3 "drawn and while holding" combat scenarios with takedowns.

Any Local Instructor Additions.
Current and subject to change test fee is $100.
...After Level 3, a basic Instructorship is $150, should one wish the additional title.

Force Necessary Stick: Level 4 Test

Eight more hours experience working on these Level 4 materials. Time and grade in similar themed systems also may count upon approval.

Knowledge and Understanding
* A working and conversational knowledge of all the *Stop 4* commonalities.
* A working and conversational knowledge of the *When* question, as it relates to batons.
* A working and conversational knowledge of *Stop 4* as it relates to impact weapon fights.

Physical Accomplishments
* The Step and Slide footwork while holding a stick, on the Combat Clock.
    - while holding a one-hand grip.
    - while holding a two-hand grip.

* Demonstrate the single-hand grip impact weapon strikes
    - in the air.
    - hitting training gear.

* Demonstrate the closed baton strikes
    - in the air.
    - hitting training gear.

* Demonstrate the Level 4 Stalk and Mad Rush exercise where the trainer stops invading at the shoulder line collision.
    * show 1 scenario with a stick quick draw and strike, maybe a takedown.
    * show 3 "drawn and while holding" combat scenarios with takedowns.

Any Local Instructor Additions.
Current and subject to change test fee is $100.

FORCE NECESSARY STICK: LEVEL 5 TEST

Eight more hours experience working on these Level 4 materials. Time and grade in similar themed systems also may count upon approval.

Knowledge and Understanding
* A working and conversational knowledge of all the *Stop 5* commonalities.
* A working and conversational knowledge of the *How* question, as it relates to batons.
* A working and conversational knowledge of *Stop 5* as it relates to impact weapon fights.

Physical Accomplishments
* The Circle Foot while holding a stick, on the Combat Clock.
    - while holding a one hand grip.
    - while holding a two-hand grip.

Explain and Demonstrate the Double-Hand Grip Strikes and Blocks
    - in the air.
    - hitting and/or blocking training gear.

An Introduction into "Pugil-Style" two-hand grip dueling.

* Demonstrate the Level 5 Stalk and Mad Rush exercise where the trainer stops his mad rush invasion at the Bear Hug collision.
    * show 1 scenario with a stick quick draw and strike, maybe a takedown.
    * show 3 "drawn and while holding" combat scenarios with takedowns.

Any Local Instructor Additions.
Current and subject to change test fee is $100.

Force Necessary Stick: Level 6 Test

Ten more hours experience working on these Level 6 materials. Time and grade in similar themed systems also may count upon, approval.

Knowledge and Understanding
* A working and conversational knowledge of all the *Stop 6* commonalities.
* A working and conversational knowledge of the *Why* question, as it relates to batons.
* A working and conversational knowledge of *Stop 6* as it relates to impact weapon fights.

Physical Accomplishments
* The Pivot Footwork while holding a stick, on the Combat Clock.
    - while holding a one hand grip.
    - while holding a two-hand grip.

* Demonstrate the Level 6 Stalk and Mad Rush Exercise where the trainer stops his mad rush invasion when the trainer and trainee have a ground collision.
    * show 1 scenarios with a stick quick draw and strike, maybe a takedown.
    * show 6 "drawn while holding" ground combat scenarios.

* Experiment with impact weapon dueling

Any Local Instructor Additions.
Current and subject to change test fee is $100.
...After Level 6, an advanced Instructorship is $150, should one wish the additional title.

Force Necessary Stick: Level 7 Test

Requirement: Perform these Takedowns in Combat Scenarios.
1: Head Takedown 1 - Any significant strike to the head.
2: Head Takedown 2 - Head Vice Takedown.
3: Neck Takedown 1 - Significant strike to the neck.
4: Neck Takedown 2 - Neck Vice Takedown.
5: Neck Takedown 3 - Rear Neck Pull Takedown.
6: Neck Takedown 4 - Fist Ram Choke Rear Pull Takedown.
7: Neck Takedown 5 - The Interlocking Choke Takedown.
8: Neck Takedown 6 - Counter the Tackle Takedown.
9: Neck Takedown 7 -The X Choke Takedown Series: The Rear.
10: Neck Takedown 8 -The X Choke Takedown Series: The Front.
11: Neck Takedown 9 -The X Choke Takedown Series: The Sides.
12: Neck Takedown 10 - The Supported Chokes Series.
13: Neck Takedown 11 - Wheel Throw on the Neck's Back.
14: Neck Takedown 12 - The Stomach Choke.

Any Local Instructor Additions.
Current and subject to change test fee is $100.

Force Necessary Stick: Level 8 Test

Requirement: Perform these Takedowns in Combat Scenarios.
- 15: Torso Takedown 1 - The Clavicle Pull Down.
- 16: Torso Takedown 2 - The Chest Pull Back.
- 17: Torso Takedown 3 - The Rear Pelvis Pull.
- 18: Torso Takedown 4 - The Front Pelvis Pull.
- 19: Torso Takedown 5 - The Riot Stick Tip Strike to Stomach.
- 20: Torso Takedown 6 - The Stomach Throw.

- 21: Arm Takedown 1 - The Biceps Lever.
- 22: Arm Takedown 2 - The Triceps Level.
- 23: Arm Takedown 3 - The Bent Arm Takedown.
- 24: Arm Takedown 4 - Arm to Neck Bridge.
- 25: Arm Takedown 5 - The Snake Killer Takedown Series.
- 26: Arm Takedown 6 - The Straight Side Arm Bar.

Any Local Instructor Additions.
Current and subject to change test fee is $100.

Force Necessary Stick: Level 9 Test

Requirement: Perform these Takedowns in Combat Scenarios.
- 27: Arm Takedown 7 - The Rear Arm Bar Hammerlock.
- 28: Arm Takedown 8 - The "Branch Down" Bent Arm Bar.
- 29: Arm Takedown 9 - The Reverse Arm Bar.
- 30: Arm Takedown 10 - The Branch Down Push or Punch Catch.
- 31: Arm Takedown 11 - Bent Arm Takedowns.

- 32: Leg Takedown 1 - The Big Catch and Roll-over.
- 33: Leg Takedown 2 - The Pelvis Pull Takedown.
- 34: Leg Takedown 3 -The Thigh Shove Takedown.
- 35: Leg Takedown 4 -The Knee Push Takedown Series.
- 36: Leg Takedown 5 -The Knee X-Pull.
- 37: Leg Takedown 6 -The Ankle X-Pull.
- 38: Leg Takedown 7 - The Fire Pole.
- 39: Leg Takedown 8 - The Lower Leg Strike (such as kneecap).

- 40: Any 3 Cane and/or Umbrella Handle Takedowns.
- 41: Any one or more takedowns you collected not covered on this list.

Any Local Instructor Additions.
Current and subject to change test fee is $100.
After Level 9, an expertise Instructorship is $150, should one wish the additional title.

FORCE NECESSARY STICK: LEVEL 10, BLACK BELT TEST

Test Task Group 1) Experience.
Experience all 9 levels of instruction and achieve over 200 hours training in this and other "stick" related, approved courses.

Test Task Group 2)
Explain the theory of "proper use of force." Discuss moral, ethical and legal implications of a fight, stick or otherwise.

Test Task Group 3) Demonstrate the Pre-Combat stances upon command.
- Parade Rest Quad.
- Parade Rest Hamstrings.
- Port Arms.
- Saber Grip Concealed.
- Reverse Grip Concealed.
- Riot Stance.

Test Task Group 4) Demonstrate the Non-Ready and Ready "Stance" Positions.
- "Bus stop" stance.
    - surprise/ambush ready.
    - unaware and not ready.
- Standing-stick forward.
- Standing-stick neutral.
- Standing-stick back.
- Knee high and the three heights.
- Down on your back.
- Down on your left, and right sides.

Test Task Group 5) Stress Quick Draws.
- 6 scenarios - Stress draw the fixed baton through the Stop 6.
- 6 scenarios - Stress draw the expandable baton though the Stop 6.
  (Note may be attacked by hand, stick, or knife).

Test Task Group 6) Demonstrate all the Combat Clock angles of attack drill used.
* Basic Clock 4.
* Advanced Clock 12.
* Be prepared to demonstrate all single hand stick strikes.
* Be prepared to demonstrate all double hand stick strikes.
* Be prepared to demonstrate all support hand strikes on pads.
* Be prepared to demonstrate all support kicks on pads.

Test Task Group 7) While Holding Freestyle on a heavy bag.
    Rounds with right hand grip.
    Rounds with left hand grip.
    Switch grips from R to L while doing so.
    Rounds with two-hand grip.

Test Task Group 8) "The Nasty 90" Stick Fights!
    Stick versus unarmed attacks.
        4 scenarios vs. thrusting hand strikes.
        4 scenarios vs. hooking hand strikes.
        4 scenarios vs. kicks.
        4 scenarios countering stick and arm grab combinations.
        4 scenarios grounded vs. topside attacker.
        2 scenarios a 3rd party rescue/fight break-up.

    Stick versus knife attacks
        6 Rounds of stick vs. knife sparring, to a winning finish, accepted by judges.
        8 scenarios vs. thrusting stabs.
            * 3 on the ground vs. standing man.
        8 scenarios vs. hooking slashes.
            * 3 on the ground vs. standing man.
        4 scenarios with impact weapon quick draws vs. knife attacks.
        4 scenarios grounded vs. topside attacker.
        2 scenarios a 3rd party rescue from a knife attacker.

    Stick versus stick attacks
        6 Rounds of stick sparring, to a winning finish, accepted by judges.
        8 scenarios vs. thrusting stick stabs.
        8 scenarios vs. hooking stick slashes.
        4 scenarios with impact weapon quick draws vs. stick attacks.
        4 scenarios grounded vs. topside attacker.
        2 scenarios a 3rd party rescue from a knife attacker.
        1 scenarios vs. a stick quick draw.
        1 scenarios a 3rd party rescue from a stick attacker.

    Bastard Mix Stick versus other
        3 scenarios vs. another surprise weapon.

Any Local Instructor Additions.
Current and subject to change test fee is $500 per black belt

FORCE NECESSARY: STICK LEVEL 11: STICK AND KNIFE STUDY
(Not included in this book but rather in Hock's Pacific Archipelago course.)

FORCE NECESSARY: STICK LEVEL 12: STICK AND STICK (DOUBLE STICK STUDY)
(Not included in this book but rather in Hock's Pacific Archipelago course.)

FORCE NECESSARY: STICK LEVEL 13, 14 AND BEYOND:
Other stick-related subjects as requested and approved by W. Hock Hochheim
(Not included in this book.)

FORCE NECESSARY: STICK INSTRUCTOR RANKS:
- Class Organizer - At any level.
- Basic Instructor - After Level 3.
- Advanced Instructor - After Level 6.
- Expertise Instructor - After Level 9.
- Black Belt Instructor - After Level 10.

ADDENDUM: THE WAR POST

Nothing has improved my stick or knife training more than hitting hard targets with real weapons. Chop down these war posts! You can afford to replace them cheaply and quickly. I made posts for about $80 to $100 through the years. While it may cost more, the expense is still low.

1: To build a single war post, you need -
  a) a cheap metal bucket.
  b) a cinder block.
  c) a fence post.
  d) a bag of cement.

2: Cut the fence post down to your required height/size.

3: Mix the cement into the bucket. Set the cinder block inside, holes up. Fit the pots and make sure to get cement into the cinder block holes.
Leave the post to dry. Support them so they will stand straight.

4: Here are single and double dried-and-set samples.

5: Attach add-on arms with some rope. Watch out for splinters!

Get the Training Mission *books and videos. They contain vital, foundational information for self defense, survival to support hand, stick, knife and gun. Plus, they follow the rank-by-rank, step-by step requirements of each course.*

Thank you to:

Jane Eden - Editor
Scott Pederson - Assistant Editor
Tobias Gibson - Contributing Editor

Mark Shuey - Special Contributor

Tom Barnhart - Training Assistant
Lynn Newby-Fraser - Training Assistant
Dean Goldade - Training Assistant
Jeff "Rawhide" Laun - Training Assistant
David "Dawg" Kerwood - Training Assistant
Tim Llacuna - Training Assistant
Tom Pierce - Training Assistant
Randy Roberson - Training Assistant

Despite the "axe handle" title, this video covers all the stick strikes and blocks, plus an introduction to grappling.

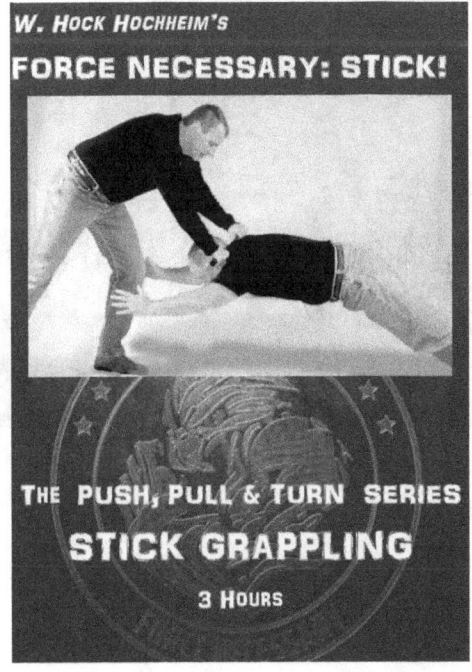

Download these training videos and more at...

www.ForceNecessary.com

www.ingramcontent.com/pod-product-compliance
Lightning Source LLC
Chambersburg PA
CBHW051803100526
44592CB00016B/2537